T0233924

Waves in Biomechanics

THz Vibrations and Modal Analysis in Proteins and
Macromolecular Structures

Synthesis Lectures on Wave Phenomena in the Physical Sciences

Editor

Sanichiro Yoshida, *Southeastern Louisiana University*

The aim of this series is to discuss the science of various waves. It consists of several books, each covering a specific subject known as a wave phenomenon. Each book is designed to be self-contained so that the reader can understand the gist of the subject. From this viewpoint, the reader can read any book as a stand-alone article. However, it is beneficial to read multiple books as it would provide the reader with the opportunity to view the same aspect of wave dynamics from different angles.

The targeted readership is graduate students and engineers whose background is similar but different from the subject. Throughout the series, it is intended to help students and engineers deepen their fundamental understanding of the subject as wave dynamics. An emphasis is laid on grasping the big picture of each subject without dealing with detailed formalism, and yet understanding the practical aspects of the subject. To this end, mathematical formulations are simplified as much as possible and applications to cutting edge research are included. The reader is encouraged to read books cited in each book for further details of the subject.

Waves in Biomechanics: THz Vibrations and Modal Analysis in Proteins and Macromolecular Structures

Domenico Scaramozzino, Giuseppe Lacidogna, and Alberto Carpinteri

ISBN: 978-3-031-01486-4   paperback
ISBN: 978-3-031-02614-0   PDF
ISBN: 978-3-031-00358-5   hardcover

DOI 10.1007/978-3-031-02614-0

A Publication in the Springer series
SYNTHESIS LECTURES ON WAVE PHENOMENA IN THE PHYSICAL SCIENCES

Lecture #4
Series Editor: Sanichiro Yoshida, *Southeastern Louisiana University*
Series ISSN
Print 2690-2346   Electronic 2690-2354

Waves in Biomechanics

THz Vibrations and Modal Analysis in Proteins and Macromolecular Structures

Domenico Scaramozzino, Giuseppe Lacidogna, and Alberto Carpinteri
Politecnico di Torino

SYNTHESIS LECTURES ON WAVE PHENOMENA IN THE PHYSICAL SCIENCES #4

ABSTRACT

Proteins and macromolecular structures represent one of the most important building blocks for a variety of biological processes. Their biological activity is performed in a dynamic fashion, hence the concepts of waves and vibrations can help to explain how proteins function. This book has the goal of highlighting the importance of wave and vibrational phenomena in the realm of proteins. It targets younger students as well as graduate researchers who work in various scientific fields and are interested in learning how mechanical vibrations affect and drive the biological activity of proteins and macromolecular structures. Great attention is given to the computational approaches dedicated to the evaluation of protein dynamics and biological behavior, and modern experimental techniques are addressed as well. The book is written in a way that non-experts in the field can grasp most of the presented subjects. However, it is also based on the most relevant and recent scientific literature, providing a rather comprehensive library for the reader eager to know more about specific topics.

KEYWORDS

protein, wave, protein dynamics, protein flexibility, protein low-frequency vibration, modal analysis, elastic network model, protein conformational change, protein hinge-domain motion, Raman spectroscopy, THz-TDS

Contents

Preface

There is a brief Latin aphorism that says *motus est vita*, meaning *motion is life*. No sentence could be better suited to describe the purpose of this short book, entitled *Waves in Biomechanics: THz Vibrations and Modal Analysis in Proteins and Macromolecular Structures*. The goal of this book is to provide a basic understanding of the motions and vibrational phenomena that affect and regulate the behavior of those tiny fundamental entities which operate at the ground level of biology, i.e., proteins and macromolecular structures. Proteins are key players in biological processes, and therefore in life processes by extension. Their job is carried out in a relentless and dynamic fashion, hence the study of protein dynamics and vibrations is pivotal for an understanding of their biological mechanisms and functions. Once again, *motus est vita*. So, this is the object of this book. We will learn what proteins are, how their dynamics can be investigated, and finally how to analyze their biological activity based on their dynamics. This book is divided into five chapters, whose main contents are as follows.

Chapter 1 provides a brief Introduction regarding the fundamentals of dynamics, vibrations, and wave propagation theory. The case of the harmonic oscillator first will be addressed to lay out the fundamental equations of motion. This will be done in the case of both the single- and multi-degree-of-freedom harmonic oscillator, with and without damping. Finally, the fundamentals of wave propagation theories will be addressed, with special focus on the concepts of frequency, wavelength, and speed of propagation.

Chapter 2 contains a specific description of proteins. First, we will start by describing the building blocks of proteins, the amino acids, and will see how these combine themselves to generate the various levels of protein structure. Then, we will learn about the phenomenon of protein folding, which enables the protein to reach its functional three-dimensional conformation through energy-minimization steps. Finally, we will discuss the fundamental paradigm adopted nowadays to study protein behavior, which relies on protein dynamics as the fundamental bridge between structure and function.

Chapter 3 addresses in more detail the problem of investigating protein dynamics from a computational perspective. After briefly describing molecular dynamics (MD) simulations, large space will be given to simplified analyses, such as Normal Mode Analysis (NMA). Afterward, simplified models for the extraction of the low-frequency protein vibrations will be described, which are based on single-parameter harmonic potentials and are known as Elastic Network Models (ENMs). Finally, Structure Mechanics-based approaches relying on elastic lattice models and finite elements, as well as other approaches based on rigid blocks, for the evaluation of protein flexibility and dynamics, will be reported.

Chapter 4 shows how the low-frequency protein vibrations and mode shapes are deeply connected to the biological mechanisms and functions through the concept of conformational change. In particular, we will see how even simplified models, such as ENMs, are able to provide fairly accurate predictions of the observed protein mechanisms. Finally, a new model based on a hinge-domain representation of the protein structure (hdANM) will be described, which is useful for the prediction and visualization of large-scale protein hinge-domain motions.

Finally, Chapter 5 illustrates the most recent findings about protein dynamics and vibrations from an experimental viewpoint. Specifically, Raman spectroscopy and THz time-domain spectroscopy (THz-TDS) will be described and their use to detect and investigate protein vibrations (especially in the low-frequency range) will be addressed.

This book is part of a series entitled "Wave Phenomena in the Physical Sciences," and has the goal of providing knowledge about wave and vibrational phenomena in the realm of proteins and macromolecular structures. We would like to express our sincere gratitude to Prof. Sanichiro Yoshida, who has kindly invited us to bring our contribution to this series with our studies in the field of biology and biomechanics. Moreover, we would also like to thank Paul Petralia of Morgan & Claypool Publishers for his support during the writing and printing phases of this book.

Domenico Scaramozzino, Giuseppe Lacidogna, and Alberto Carpinteri
October 2021

C H A P T E R   1

Introduction

What is a vibration? A vibration can be defined as a dynamical state in which a system undergoes a certain motion about its equilibrium position and whose time-history depends on the system characteristics and external excitations. In our everyday life, we can observe all sort of vibrations, such as periodic vibrations, random vibrations, decaying vibrations, etc. The investigation of vibrations allows us to understand the intrinsic behavior of the system and is carried out in a variety of fields and applications, e.g., in mechanical engineering for the analysis of turbines and rotors, in civil engineering for the monitoring of buildings and bridges, etc.

Vibrations are also closely related with waves. Waves are the result of the propagation of a certain disturbance in the time- and space-domain. Waves can be generated by vibrations and they can also induce vibrations in a certain system. In this Introduction, we will discuss the basics of vibrations and waves, starting from the simple single-degree-of-freedom (SDOF) harmonic oscillator.

1.1   THE HARMONIC OSCILLATOR

Consider the basic system reported in Figure 1.1. A body with mass m is connected to a rigid support via a spring with stiffness k and a damper with damping c. The system has only one degree of freedom (DOF), which is represented by the horizontal displacement δ. The equilibrium between the inertia force F_i, the elastic force F_e, and damping force F_d yields:

$$F_i + F_d + F_e = m\ddot{\delta} + c\dot{\delta} + k\delta = 0, \tag{1.1}$$

where $\ddot{\delta}$ represents the acceleration of the body and $\dot{\delta}$ its velocity. Equation (1.1) can be rewritten by dividing all members by m as:

$$\ddot{\delta} + 2\xi\omega\dot{\delta} + \omega^2\delta = 0, \tag{1.2}$$

where $\omega^2 = k/m$ and $\xi = c/2m\omega$. ω represents the angular frequency of vibration (rad/s) and ξ is the dimensionless damping coefficient. The solution of the second-order differential equation reported in Equation (1.2) can be looked for in the general form $\delta(t) = Ae^{\lambda t}$. Substitution of this solution into Equation (1.2) yields:

$$\lambda^2 + 2\xi\omega\lambda + \omega^2 = 0 \;\rightarrow\; \lambda_{1,2} = -\xi\omega \pm \omega\sqrt{\xi^2 - 1}. \tag{1.3}$$

Equation (1.3) has four different solutions depending on the value of the damping coefficient ξ, i.e., for $\xi = 0$, $0 < \xi < 1$, $\xi = 1$, and $\xi > 1$. In the first case, we have no damping and

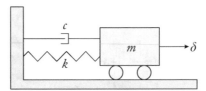

Figure 1.1: Sketch of the SDOF harmonic oscillator, including the term of mass, stiffness, and damping.

therefore we talk about undamped oscillations. In the second case, the damping coefficient is sufficiently small that we observe the so-called underdamped oscillations. In the third case, we have the critical damping condition. In the fourth case, we have overdamped vibrations. Based on the value of ξ, the function $\delta(t)$ has different mathematical expressions. Moreover, since Equation (1.2) is a second-order differential equation, we need two conditions to completely set up the function $\delta(t)$. These conditions are usually referred to the initial state ($t = 0$), i.e., $\delta(0) = \delta_0$ and $\dot{\delta}(0) = \dot{\delta}_0$. Below, the expressions of $\delta(t)$ are reported for the four cases:

(a) $\xi = 0$: $\lambda_1 = i\omega$ and $\lambda_2 = -i\omega$, being $i^2 = -1$. In this case we have periodic harmonic oscillations with angular frequency ω and period $T = 2\pi/\omega$:

$$\delta(t) = \delta_0 \cos(\omega t) + \frac{\dot{\delta}_0}{\omega} \sin(\omega t). \tag{1.4}$$

(b) $0 < \xi < 1$: $\lambda_1 = -\xi\omega + i\omega_d$ and $\lambda_2 = -\xi\omega - i\omega_d$, being $\omega_d^2 = \omega^2(1 - \xi^2)$ the so-called reduced frequency. We have now oscillations with frequency ω_d, exponentially decaying over time due to damping:

$$\delta(t) = e^{-\xi\omega t}\left[\delta_0 \cos(\omega_d t) + \frac{\dot{\delta}_0 + \xi\omega\delta_0}{\omega_d} \sin(\omega_d t)\right]. \tag{1.5}$$

(c) $\xi = 1$: $\lambda_1 = \lambda_2 = -\omega$. In this case we have the critical damping condition where no more continuous oscillations are detected:

$$\delta(t) = e^{-\omega t}\left[\delta_0 + \left(\dot{\delta}_0 + \omega\delta_0\right)t\right]. \tag{1.6}$$

(d) $\xi > 1$: $\lambda_1 = -\xi\omega + \omega(\xi^2 - 1)^{1/2}$ and $\lambda_2 = -\xi\omega - \omega(\xi^2 - 1)^{1/2}$. Here we are in the overdamped limit and no oscillations are detected as ω_d becomes imaginary:

$$\delta(t) = \frac{e^{-\xi\omega t}}{2}\left\{\left[\delta_0 + \frac{(\dot{\delta}_0 + \xi\omega\delta_0)}{\omega\sqrt{\xi^2 - 1}}\right]e^{\omega\sqrt{\xi^2 - 1}t}\right.$$
$$\left. + \left[\delta_0 - \frac{(\dot{\delta}_0 + \xi\omega\delta_0)}{\omega\sqrt{\xi^2 - 1}}\right]e^{-\omega\sqrt{\xi^2 - 1}t}\right\}. \tag{1.7}$$

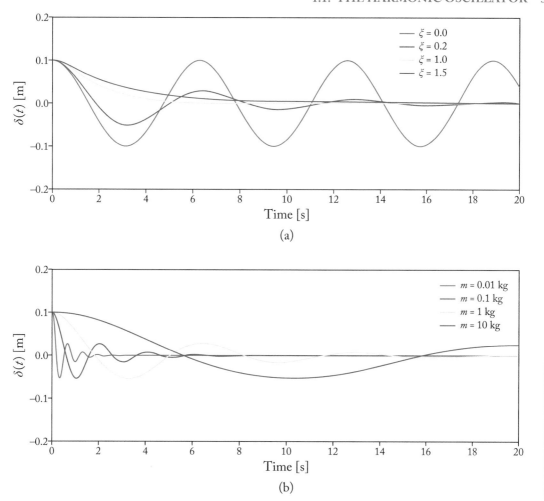

Figure 1.2: Response of a SDOF oscillator with $k = 1$ N/m, $\delta\,(0) = 0.1$ m, $\dot\delta(0) = 0$ m/s: (a) fixed mass $m = 1$ kg and varying $\xi = 0, 0.2, 1$, and 1.5; (b) fixed $\xi = 0.2$, and varying mass $m = 0.01, 0.1, 1$, and 10 kg.

Figure 1.2a shows the time-evolution of $\delta(t)$ for a system with $m = 1$ kg and $k = 1$ N/m, $\delta(0) = 0.1$ m, $\dot\delta(0) = 0$ m/s, and four values of ξ (0, 0.2, 1, and 1.5). Increasing the damping coefficient leads to a decrease of the dynamical evolution over time. Figure 1.2b shows the same system with $k = 1$ N/m, $\xi = 0.2$, $\delta(0) = 0.1$ m, $\dot\delta(0) = 0$ m/s, and four values of m (0.01, 0.1, 1, and 10 kg). As can be seen, increasing the mass m decreases the vibrational frequency ω.

So far, we have seen how to investigate the free vibrations of an SDOF system. What happens if the system we are investigating has more than one DOF? In this case, we are dealing with a multi-degree-of-freedom (MDOF) problem. Equation (1.1) can be rewritten in a formally analogous manner for an MDOF system, by replacing the scalar quantities with the vector and matrix counterparts, i.e.,

$$F_i + F_d + F_e = M\ddot{\delta} + C\dot{\delta} + K\delta = 0, \tag{1.8}$$

where F_i, F_d, and F_e are vectors of inertia, damping, and elastic forces, which depend, respectively, on the mass matrix M, the damping matrix C, and the stiffness matrix K, and the vectors of accelerations $\ddot{\delta}$, velocities $\dot{\delta}$, and displacements δ, which are associated with the DOFs of the system. Note that all the matrices are square matrices with dimension $N \times N$, being N the total number of DOFs, and all the vectors are column vectors with dimension $N \times 1$.

If the damping matrix C is zero, we are dealing with the classical problem of the MDOF free-vibration analysis, which allows to compute the natural vibrational frequencies of the system as well as its mode shapes. Specifically, looking for a harmonic solution of the form $\delta(t) = A\,e^{i\omega t}$, one obtains:

$$\left(K - \omega^2 M\right) A\,e^{i\omega t} = 0. \tag{1.9}$$

Equation (1.9) leads to the non-trivial solution, i.e., $A \neq 0$, only when the determinant of the matrix within the parenthesis is equal to zero. This leads to an N-order polynomial, which characterizes the so-called eigenvalue-eigenvector problem. The N roots of the polynomial are the eigenvalues, which represent the values of the natural angular frequencies ω_n, while the corresponding N eigenvectors represent the mode shapes A_n, which describe the spatial motion associated with the nth mode [1].

1.2   WAVES

As we briefly mentioned above, vibrations can generate waves and waves can induce vibrations. Consider, for a moment, a guitar string. When a perturbation is applied to the string, such as the one induced by fingers, the spring generally undergoes underdamped vibrations. That is, the spring oscillates for a bit about its equilibrium position, with oscillations getting smaller and smaller over time due to damping. During the motion of the spring, the surrounding air molecules start vibrating according to the motion of the spring. This motion then propagates throughout the environment, as the air molecules which were closer to the spring start transferring their kinetic energy to the ones lying farther. This phenomenon is what we perceive as sound, and it is nothing more than an acoustic wave generated by the vibration of the air molecules and initiated by the vibration of the guitar spring. When we pinch the guitar spring in a different way, we are changing its vibrational mode and therefore its frequency. This will, in turn, affect the frequency of vibration of the air molecules, and we will hear a different sound, resulting in a different musical note.

Waves are therefore oscillations which propagate both in time and space. Accordingly, their evolution is described by two fundamental parameters. The first is the angular frequency in the time-domain ω, which measures how many radians the wave accomplishes in the unit time. Note that the angular frequency ω is related to the vibrational frequency f by a factor of 2π ($\omega = 2\pi f$). The second is the angular frequency in the space-domain k', also called wavenumber, which measures how many cycles the wave accomplishes in the unit space (the apex $'$ has been inserted here not to confuse k' with the k reported above, which was referred to the stiffness of the spring in Figure 1.1). Note that the inverse of the wave number is proportional to the wavelength ψ ($\psi = 2\pi/k'$), which is a measure of the space distance between two subsequent points with the same amplitude value.

If we consider a simple wave traveling in one direction along the x-axis without damping, we can describe the evolution of its amplitude Ψ in both the time- and space-domain as follows:

$$\Psi = \Psi_0 \sin(\omega t \pm k'x), \tag{1.10}$$

where Ψ_0 represents the maximum amplitude of the wave and the sign $\pm$ has been inserted to take into account that the wave travels both in the positive and negative direction of the x-axis [2]. Based on the value of ω and k' (or similarly f and ψ), we can define the wave velocity v as the ratio of the two:

$$v = \frac{\omega}{k'} = \frac{2\pi f}{2\pi/\psi} = f\psi. \tag{1.11}$$

The velocity v accounts for the speed of propagation of the wave and it depends on the characteristics of the wave (longitudinal wave, transverse wave, etc.) and on the propagation medium [2]. For example, acoustic waves can propagate in air, liquids, and solids and their velocity varies accordingly. In air, they travel at a speed of about 300 m/s, whereas they travel at about 1500 m/s in water and at 5000 m/s in rigid solids such as steel. Conversely, electromagnetic waves travel in vacuum at the speed of light (3×10^8 m/s), while their speed decreases when traveling in a medium. More details about the fundamentals of waves, including equations, basic properties, and their propagation can be found in Yoshida [2].

C H A P T E R   2

Proteins: The Basis of Biological Mechanisms

Proteins are one of the most important building blocks for the achievement of the numerous biological reactions that occur every day in our body. The oxygen that we introduce into our lungs while we are breathing is successfully delivered to tissues and organs by the relentless job of hemoglobin, a protein contained in our red blood cells. The biochemical process which allows our eyes to interpret the images of the real world is mediated by proteins in charge of facilitating the conversion of photons into electrochemical signals which are then sent to the brain for final interpretation. Other proteins, such as collagen, are extremely important to provide specific parts of our body adequate levels of stiffness and mechanical resistance. Proteins like kinesin are essential for the transportation of nutrients throughout the cell environment. The function of transmembrane proteins, which are embedded within the cellular membrane, is crucial for regulating the movement of ions, nutrients and small molecules to and from the cell. Proteins also act as enzymes, catalyzing biochemical reactions and improving the reaction rate by lowering the activation energy. Other classes of proteins, the antibodies, are extremely important to fight against and neutralize viruses and bacteria.

As can be seen from the examples reported above, the role of proteins can be very diverse. All these activities contribute to the correct functioning of our body as a whole. Shortly and in a non-exhaustive way, we can summarize the main tasks and functions performed by the proteins in the human body, as shown in Figure 2.1:

- transportation, both throughout the body (e.g., hemoglobin), within the cell (e.g., molecular motors, such as kinesin), and across the cell (e.g., transmembrane proteins);

- defense, mostly explicated by our antibodies;

- structural and mechanical function, e.g., providing the required stiffness to ligaments and tendons (e.g., collagen), or participating in the process of muscle contraction (e.g., actin);

- enzymatic and catalytic function (e.g., lysozyme, which is in charge of breaking internal bonds in peptidoglycans); and

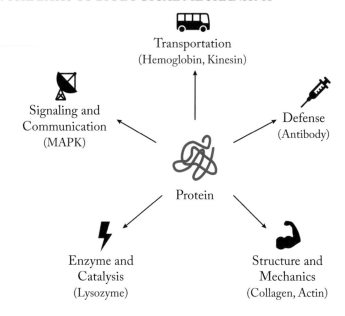

Figure 2.1: Some fundamental protein tasks and functions in the human body.

- signaling and communication (e.g., proteins such as mitogen-activated protein kinase, MAPK, which are involved into signaling pathways needed for the regulation of specific cell stimuli).

Regardless of the different tasks performed by the specific protein, each is made up of the same basic building block: the amino acid. Ultimately, what really differentiates the various protein functions only depends on the different combination and arrangement of these tiny building blocks that leads to the generation of a variety of structures. When dealing with protein structures, four different hierarchical levels are usually distinguished, ranging from the primary structure up to the quaternary structure. In the following paragraphs, we will recount in more details the fundamental features of the four protein structural levels, how their creation takes place, and how such a process is connected to the protein functionality. More exhaustive details about the biological, biochemical, and biophysical properties of proteins are thoroughly described in [3, 4].

2.1 FROM THE SINGLE AMINO ACID TO THE THREE-DIMENSIONAL PROTEIN STRUCTURE

As briefly reported above, the basic building block of every protein is the amino acid. The amino acid is a chemical group, a simple sketch of which is shown in Figure 2.2. This chemical group is made up of a central carbon atom, usually called alpha-carbon (C^α), which is covalently bonded

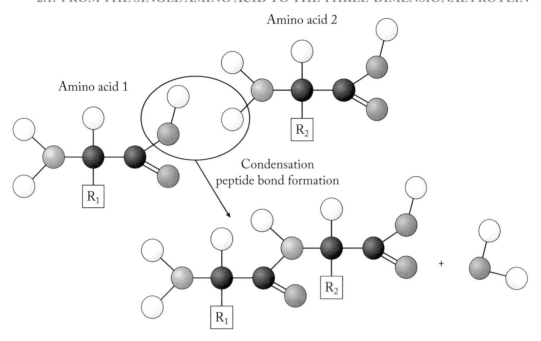

Figure 2.2: Structure of the basic amino acid and condensation reaction between two amino acids. After the condensation reaction for the creation of a dipeptide, a water molecule is released. Colors for the atoms: white for hydrogen, black for carbon, red for oxygen, and cyan for nitrogen.

to an amine group ($-NH_2$), a carboxyl group ($-COOH$), a hydrogen atom (H), and a side chain R_i. The side chain R_i is what distinguishes the different amino acids from each other. In the cell and body environment, these molecules usually exist in ionized form, so that the amine group can be often found in the $-NH_{3+}$ form and the carboxyl group in the $-COO^-$ form.

Different amino acids can interact between each other. In this case, when the carboxyl group of one amino acid reacts with the amine group of another, the two amino acids undergo a chemical reaction where a peptide bond is generated. This bond is very strong and is formed between the carbon atom of the carboxyl group and the nitrogen atom of the amine group. This phenomenon is known as condensation reaction, due to the fact that a water molecule is released as a secondary product. As a result of the condensation between two amino acids, a dipeptide is then produced. A sketchy representation of the condensation between two amino acids is reported in Figure 2.2.

This process is often repeated among more than two amino acids. In this case, we obtain a chain of several amino acids, which is called a polypeptide. This polymerization process is what ultimately creates proteins, which are nothing more than a chain of amino acids. Note that, in

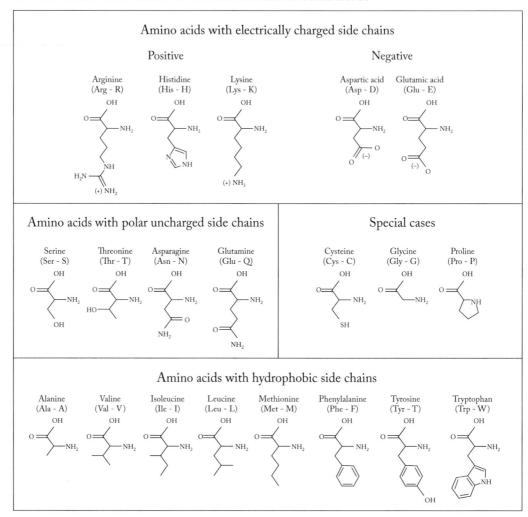

Figure 2.3: The most common twenty amino acids found in eukaryotes, classified depending on their hydrophobic-hydrophilic characteristics and their charge. Other than the names, the conventional three-letter and one-letter codes are also reported for each amino acid.

proteins, the amino acids are often referred to as "residues," because this is what remains when they join together into the polypeptide chain after the release of water molecules.

We have seen that what ultimately makes the amino acids different between each other is their side chain R_i. This side chemical group can differ in structure, polarity and electrical charge (Figure 2.3). Hence, depending on the features of the side chain, we can distinguish between polar amino acids, nonpolar amino acids and electrically (positive or negative) charged amino

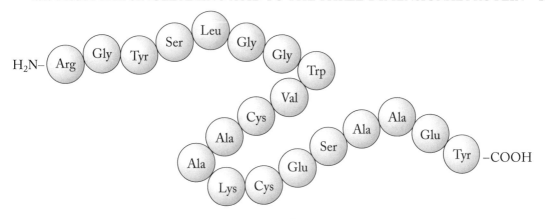

Figure 2.4: Protein primary structure: example of a short 20-amino acid polypeptide. Beads indicate amino acids according to their three-letter alphabetical code.

acids. Charged and polar side chains have affinity with water molecules, and for this reason they are called hydrophilic residues. Conversely, nonpolar side chains are hydrophobic, because they usually avoid the contact with water molecules when they arrange themselves within the protein structure. As we will see when describing the protein tertiary structure and the process of protein folding, the hydrophobicity and hydrophilicity properties of the amino acids are major key players in the generation of the final protein structure.

Under standard biological conditions (e.g., at neutral pH), some amino acids act as bases, their side chains being positively charged, whereas other residues act as acids, their side chains being negatively charged. Depending on their structure, we can also find aliphatic or aromatic groups in the side chains of amino acids.

By means of the polymerization process described above, several amino acids combine together to form a protein chain. The linear sequence of amino acids in a protein is what is called the primary structure (see Figure 2.4). Conventionally, the sequence is reported from left to right, starting from the amino-terminal of the first amino acid to the carboxyl-terminal of the last one. For this reason, the former is called the N-terminus, whereas the latter is known as the C-terminus. Each amino acid is assigned a conventional alphabetical code, therefore the primary structure can be distinguished as a specific string of letters associated with the specific amino acids included into the sequence. These alphabetical codes can be either formed by three or one letter per amino acid (see Figure 2.3).

As the protein chain gets formed through the continuous addition of more amino acids, the local interactions between their side chains start to generate well-defined geometrical segments, which are referred to as the secondary structures. The two most common secondary structures that are found in proteins are the so-called α-helix and β-sheet (Figure 2.5) [5].

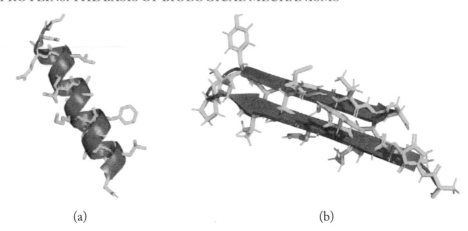

(a)   (b)

Figure 2.5: Common secondary structures in proteins: (a) α-helix and (b) β-sheet. In red, the cartoon representation of the secondary structure; in green, the sticks of the amino acid side chains.

The α-helix is formed when a segment of the protein backbone is arranged around its axis through a spiral configuration (Figure 2.5a) [6]. In protein α-helices the geometrical characteristics of the spiral are usually well-conserved, as every helical turn always counts 3.6 amino acids and the distance between two subsequent turns is equal to 5.4 Å. This secondary structure is stabilized by the hydrogen bonds generated between the carbonyl oxygen of a certain amino acid and the amide hydrogen of the amino acid which is four positions ahead in the primary structure. The α-helix is usually densely packed and the side chains of the amino acids point toward the external sides of the helix. The β-sheets are also very common secondary structures and are stabilized by a pattern of hydrogen bonds [7]. Unlike the α-helices, the geometry of β-sheets is characterized by two or more strands of amino acids connected side-by-side (Figure 2.5b). Depending on the specific pattern of the hydrogen bonding, β-sheets can be parallel or anti-parallel, and the side chains of the residues are often found out of the sheet plane.

Although not so common, other secondary structures other than α-helices and β-sheets can also be found in proteins. Examples are the π-helix, the 3_{10}-helix, and local motifs such as the end-cappings which are sometimes present at the N-termini. Loops and β-turns are also common local segments found in protein structures. We can also find what is usually called the super-secondary structures or structural motifs. These are made up of a combination of secondary structures, which create peculiar geometrical patterns. The most common examples of super-secondary structures are the β-helix, where several β-sheets are arranged together forming a helical shape, and the coiled coil, where two or more α-helices are coiled together.

The next level in the structural hierarchy of a protein, after the primary and secondary structure, is what is called the tertiary structure. It basically refers to the three-dimensional

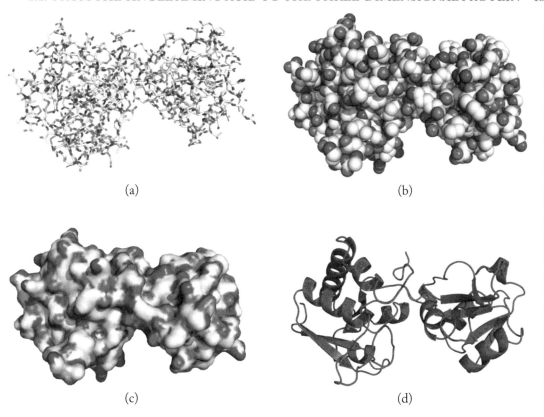

Figure 2.6: Representations of the three-dimensional structure of LAO-binding protein (PDB: 2lao): (a) sticks, (b) spheres, (c) surface, and (d) cartoon. Colors in (a–c) reflect the atom type.

geometry of the protein, which is formed and stabilized through the interactions between the various amino acids. As we will see in the remaining of this chapter and in the following ones, the tertiary structure is one of the most important features for the understanding of protein behavior and functionality. As a matter of fact, a strong correlation has been discovered between the protein three-dimensional structure and its biological mechanisms.

The tertiary structure of a certain protein can be illustrated in various graphical ways. Some examples are shown in Figure 2.6 for the lysine-arginine-ornithine (LAO)-binding protein. In Figure 2.6a the structure is represented by means of sticks, that enable to visualize all the side chains of the amino acids. Sphere representation is used in Figure 2.6b, which is convenient to have a clear idea about the volumetric occupancy and packing density. Figure 2.6c makes use of the surface visualization, where the generation of the external surface can be extremely useful for the identification of binding sites, channels and cavities. Finally, the Figure 2.6d shows the cartoon representation, which is a common way to visualize the overall shape of the protein as it enables to clearly identify the protein secondary structures.

Finally, there are certain proteins that present a fourth level of structural hierarchy, which is called the quaternary structure. With this term we mainly refer to those proteins which have more than one amino acid chain, interacting and stabilizing with each other to form the whole protein. Consequently, these proteins usually have two or more subunits. One famous example is hemoglobin, which is made of four subunits, that work cooperatively to bind more than one oxygen molecule at a time to deliver it efficiently throughout the body. The inter-subunits bonds that are required to stabilize the quaternary structure are usually weaker than the intra-subunit ones.

2.2   PROTEIN FOLDING: FROM SEQUENCE TO STRUCTURE

In the previous paragraph we surveyed the different levels of the protein structure, ranging from the elementary amino acid to the final three-dimensional shape. These levels reflect the complex process that leads from the linear amino acid sequence to the formation of the complex native protein structure, the so-called protein folding. Protein folding is a multifaceted, chemo-physical phenomenon through which the protein reaches its final functional conformation. Before describing in more details the phenomenon of protein folding, it is useful to firstly briefly describe how the amino acid chain gets formed within the cell. This phenomenon is what is called protein biosynthesis [3, 4], a simplified sketch of which is shown in Figure 2.7.

Protein biosynthesis takes place initially within the nucleus of the cell and involves the DNA and a particular enzyme, which is called RNA polymerase. The genes encoded in the DNA contain all the instructions for the fabrication of specific proteins. The transcription is the first step in protein biosynthesis and takes place when the RNA polymerase interacts with the gene sequence. The RNA polymerase attaches to the starting point of the gene and moves along it generating a strand of messenger RNA (mRNA), using all the free bases contained in the cell nucleus. The sequence of the generated mRNA strand depends on the specific coding information contained in the originating gene. At the end of the transcription, the gene information has been completely read by the RNA polymerase and transcript into the mRNA, which will then be used for the generation of the protein sequence.

At this point, when the mRNA is ready to be processed, the second step takes place outside the cell nucleus, within the cytoplasm, and is called translation. The translation involves the binding of the mRNA to a complex molecular machine, the ribosome, which is in charge of reading the code in the mRNA strand and translating it into the amino acid sequence. This is made possible by means of another key player, the transfer RNA (tRNA), which is a molecule that delivers the specific amino acids to the ribosome. Three bases of the mRNA sequence are read at a time and the tRNA delivers one corresponding amino acid to the ribosome which adds it

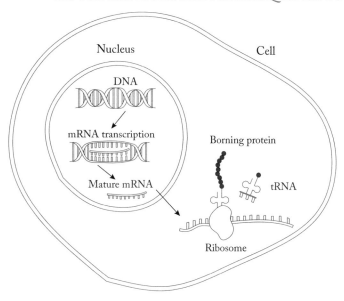

Figure 2.7: The generation of a protein (protein biosynthesis): from the DNA inside the cell nucleus to the final native conformation in the cytoplasm.

to the growing polypeptide chain. As soon as the amino acid chain goes out of the ribosome, the folding process starts, eventually leading to the native three-dimensional shape of the protein.

The folding process starts as soon as the N-terminus of the primary structure gets out from the ribosome into the cytoplasm. The secondary structures are often generated first, and then the tertiary structure of the protein is completed when the polypeptide chain completely folds upon itself. This mechanism is driven by the chemo-physical interactions that take place among the different amino acids. The native conformation of the protein is thus the final result of the complex dynamic equilibrium between the different forces acting within the system. Notice that the folding process does not occur in vacuum, but it takes place within the cell environment. Therefore, this process is strongly influenced not only by the interactions among the residues, but also by the interactions of the amino acids with all the other molecules and ions that exist within the cell.

We saw in the previous paragraphs that hydrogen bond patterns play an important role in stabilizing secondary structures, such as α-helices and β-sheets. In the same way, these interactions are pivotal for the definition of the folded shape of the protein, especially due to the fact that the protein folding occurs in an environment where there is plenty of water molecules. As a matter of fact, it is quite well known that exposition to water molecules plays an extremely important role in driving the folding process. It has been suggested that the minimization of the contacts between hydrophobic amino acids and water molecules is one of the main driving

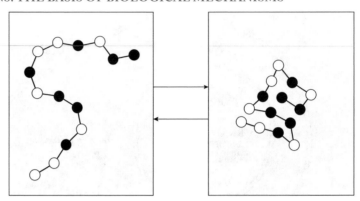

Figure 2.8: Minimizing the contacts between water molecules and hydrophobic residues is one of the main driving forces of protein folding. Left panel: unfolded protein; right panel: folded protein. Black dots: hydrophobic residues; white dots: polar residues.

forces of the process. As a consequence, the hydrophobic amino acids get buried mostly within the core of the folded protein, whereas the hydrophilic side chains are left exposed on the surface (Figure 2.8) [8]. A densely packed structure is then usually reached, which makes the folding phenomenon a thermodynamically favored mechanism, as it is carried out by minimizing the free energy of the system.

Although hydrophobic-hydrophilic interactions are one of the key factors in driving protein folding, other forces are believed to play important roles such as the Van der Wall forces. Furthermore, this phenomenon is very sensible to the characteristics of the surrounding environment, e.g., temperature, pH, ion concentration, that might lead to the protein denaturation if certain unfavorable conditions are met. For example, a correctly folded protein in the standard physiological environment might undergo denaturation when the temperature increases or the pH decreases.

Sometimes, a protein which is not able to reach the correct folded shape under certain environmental conditions can be helped by particular proteins, called molecular chaperons. These support specific proteins to prevent denaturation or to reach the correct fold. As we will see in the remainder of the book, since the folded shape of the protein is strictly related to its functionality, the correct folding is crucial for the positive fulfillment of the protein task.

It is believed that the native structure of a certain protein in standard physiological conditions is only determined by its amino acid sequence. This is what we refer to as the Anfinsen's dogma, or the thermodynamic hypothesis [9, 10]. This means that, at the conditions at which the folding phenomenon can take place correctly, the native structure of the protein is unique, stable and kinetically accessible. These concepts (uniqueness, stability, and kinetical accessibility) are related to the folding thermodynamics.

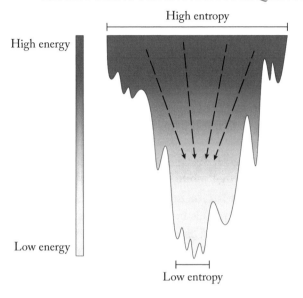

Figure 2.9: Two-dimensional representation of the protein-folding energy funnel.

At this point, a question arises that plenty of scientists have been trying to answer in the 1960s. If, for a certain amino acid sequence only one stable and kinetically accessible native structure exists, how is the polypeptide chain able to sample the entire conformational space in order to find the final correct fold? This is also connected to the Levinthal's paradox [11, 12], which states that, given the exceedingly high number of available conformations, if the final fold had to be reached by randomly sampling the whole conformational space, even if the sampling were addressed at the scale of picoseconds, the process would require more than the life of the universe even for a small polypeptide chain. Based on this observation and the Anfinsen's dogma, scientists in the 1970s established that the folding process does not proceed by random sampling, but it advances through a series of energetically favorable intermediate states.

This concept is straightforwardly understandable if you make use of the so-called protein folding energy funnel (Figure 2.9) [13, 14]. The energy funnel is a graphic representation which shows the energy level of a certain protein conformation with respect to its conformational entropy. At the top of the funnel we find the protein in its random coil conformation. Due to the geometrical and energetical properties of the random coil, in this region the protein exhibits large conformational entropy (width of the funnel) and large values of its free energy (height of the funnel). The former is due to the higher number of conformations available for the random coil, the latter is due to the thermodynamic unfavorability of this conformation. Conversely, in the lower side of the funnel we find the native folded state, which is the most stable one. Here we have the lowest energy, associated with lower values of the conformational entropy.

Based on the funnel representation, the folding phenomenon can be explained as the process which drives the polypeptide chain from the random coil configuration (top of the funnel) down to the native state (well of the funnel) through a series of intermediate states. The downward advancement throughout the funnel is what is called the folding pathway (dotted arrows in Figure 2.9). As can be seen from Figure 2.9, the energy landscape of the funnel might be quite rough and can also exhibit local minima, at high-, mid-, and low-energy values, where the protein is not in the native form yet, but it finds a stable conformation under the current environmental conditions.

In some cases, it is known that the native state of a protein can be associated with multiple conformations, since the funnel might exhibit more than a single well in its lowest region. At first sight, this might crash with the uniqueness hypothesis from the Anfinsen's dogma, but as we will see in the remainder of this book, these different configurations usually correspond to different conditions of the external environment. In several cases, the same protein can therefore exhibit multiple stable conformations depending on external factors. The typical example is the presence or absence of ligands, which induces the protein to undergo a conformational change and show different conformations. Nevertheless, these phenomena do not violate the Anfinsen's dogma.

So far we have also seen that, for a given amino acid sequence, we can obtain a specific native fold, which we know to be stable under certain environmental conditions. This means that, as soon as the environmental conditions are subjected to changes, the stability of the folded structure might be compromised. Based on the thermodynamic hypotheses, what happens here is that changes of the external conditions have the ability to modify the protein energy funnel. The funnel can then be modified both in height, width, and overall shape, leading to new and/or different local/global minima. Accordingly, this has the potential to generate new stable conformations under the different external conditions. Examples of factors that can cause a change in the energy funnel are the variation of temperature, the increase or decrease in the environmental pH, the ion concentration, the electric charge, the presence of denaturants and specific molecules, the application of forces on the protein, etc. All these factors have the capability to modify the energy landscape and alter the stability of the protein conformations.

The unfolding process can then be explained as that phenomenon leading to the change in the protein conformation due to the new external conditions. Such a process can be consequently induced, e.g., by changing the temperature or the pH of the environment, as well as by adding molecules acting as denaturants. In the latter case, we talk about chemical unfolding. Proteins might also undergo the so-called mechanical unfolding, when the forces applied to the protein get strong enough in a way that the random coil becomes the most energetically stable conformation [15].

The prediction of the folded structure for the given amino acid sequence has been one of the most challenging aspects about protein folding, and in part it still remains today. As we have seen, protein folding is a complex phenomenon that involves both chemical and physical interactions among the residues and with the specific environmental conditions. Researchers and scientists have been struggling for decades trying to find accurate methods to predict the folded shape of the protein and extensive theoretical research has been carried out for the purpose [16–18].

Sophisticated numerical models have also been developed, often based on Molecular Dynamics (MD) simulations. Although good results can be obtained for small polypeptides, for larger amino acid chains the computational costs of these models might not be practical. To overcome this computational limit, various strategies have been proposed by the scientific community in recent years. One is the use of supercomputers and distributed computing projects, as in the Folding@Home project [19], where the computational resources of volunteers worldwide are exploited to run the folding simulation. Other strategies have tried to reduce the computational cost of the problem, e.g., by developing coarse-grained models. Other modeling and theoretical strategies have also been proposed both to increase the accuracy of the predicted folded structure and minimize the computational efforts. Accordingly, every two years the international CASP (Critical Assessment of protein Structure Prediction) competition takes place, where scientists and researchers have the chance to test the accuracy of newly developed folding prediction methods.

To conclude this brief description about protein folding, it needs to be reminded that the three-dimensional structure of the protein is strictly related to the protein mechanisms and functionality. We will see that the bridge connecting the protein structure to its functionality has been related to the protein dynamics. However, when the protein is not folded properly, its tasks might not be carried out correctly. This is the case of the so-called protein misfolding. Protein misfolding might occur due to several causes, such as non-physiological environmental conditions, mutations in the amino acid sequence, and other external factors. The misfolding of certain proteins might sometimes also activate the misfolding of other proteins, leading to cascade events. This cascade might cause accumulations and aggregations of several misfolded proteins, which finally leads to the formation of amyloids. These have been shown to be associated with various neurodegenerative diseases, such as Alzheimer's [20]. Therefore, it is clear that the protein misfolding not only prevents the protein to perform its fundamental tasks, but it might also trigger the onset of harmful diseases.

2.3   FUNDAMENTAL PARADIGM OF PROTEIN ACTION: SEQUENCE-STRUCTURE-DYNAMICS-FUNCTION

The Anfinsen's dogma teaches us that each amino acid sequence leads to a well-defined folded structure, which is in turn related to protein function. Consequently, the functionality of a certain protein should already be encoded within its amino acid sequence. For this reason, the phenomenon of protein folding might also be seen as the process that allows the amino acid sequence to achieve the protein functionality through the generation of the native folded structure. This has led in the past decades to the sequence-structure-function paradigm. This paradigm was

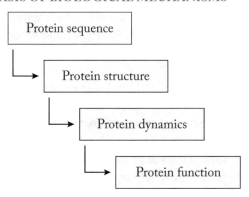

Figure 2.10: Protein sequence-structure-dynamics-function fundamental paradigm.

proposed to ultimately explain the fundamental reasons for the protein functionality and also to clarify why mutations in the amino acid sequence often lead to deleterious alterations in the protein activity.

However, while the amino acid sequence drives the formation of the three-dimensional native structure, how is the structure capable to actively influence and drive the protein behavior? One of the answers that researchers and scientists have provided to this question has been based on protein dynamics. As a matter of fact, protein dynamics has been thought to be the ultimate bridge to connect the protein structure to its functionality, thus leading to the new sequence-structure-dynamics-function paradigm (Figure 2.10).

As a matter of fact, proteins are not static in nature. At the physiological temperature, they move and vibrate at very high frequencies. The native structure is not a motionless entity, stuck in the low-energy well of the energy funnel, but it actually keeps jiggling around its equilibrium position. The specific dynamical behavior and characteristics of the protein structure is thus believed to be responsible for the protein mechanism, ultimately driving its action (Figure 2.10).

Summarizing, we can say that: (i) the amino acid sequence drives the generation of the three-dimensional structure via the folding process (as described in the previous section); (ii) the protein structure obviously influences its dynamics, as this depends on the characteristics of the structure itself; (iii) the dynamics is related to the protein functionality through the vibrational modes, which are found to reflect biologically relevant motions. Further details about the (ii) structure-dynamics and (iii) dynamics-function relationships will be provided in the following chapters.

CHAPTER 3

CHAPTER 3

Protein Vibrations and Elastic Network Models

In the previous chapter, we saw that protein dynamics is a major candidate to explain the structure-function relationship. Experimental techniques, such as X-ray crystallography and Nuclear Magnetic Resonance (NMR), have become very powerful nowadays in providing accurate descriptions of the three-dimensional protein structure. Plenty of structures are in fact resolved every year and deposited into public databases, such as the Protein Data Bank (PDB) [21]. As of December 2020, the PDB comprises a total of about 160,000 protein structures, 89% of which obtained from X-ray crystallography, 7% from NMR, and the remaining 4% resolved with other methodologies such as Electron Microscopy (EM) and hybrid methods [22]. The knowledge of the three-dimensional protein structure is obviously an important starting point in order to investigate its mechanism of action. A thorough analysis of its dynamics subsequently allows to predict its functional behavior. Hence, how can we investigate protein dynamics?

One of the most used approach is through Molecular Dynamics (MD) simulations [23, 24]. MD is based on solving the Newton's laws of motion for a system subjected to certain forces. The initial model of the system is usually taken from experimental structures, which can be represented at various levels of detail. Once the system is built, the forces acting on every atom are computed based on the derivation of potential energies. Different forms of potentials can be used for this purpose. Often these potential energies include spring terms for bond stretching and angle distortions, Lennard-Jones terms for long-range interactions, as well as Coulomb's laws for Van der Waals and electrostatic effects [24]. Obtained the forces acting on the systems, the equations of motion are integrated to obtain the velocities and positions of all the atoms. This is carried out by adopting numerical integration strategies, which require discretization into small time frames of the order of femtoseconds. Although MD simulations can provide detailed and often accurate predictions of the trajectories of the system, the high computational burden and the difficulty to achieve stable solutions at very large time scales often constitute serious obstacles.

The first MD protein simulation was reported in 1977 by McCammon et al. [25], who investigated the dynamics of a small globular protein, bovine pancreatic trypsin inhibitor (BPTI). The simulation, which lasted 8.8 ps and was carried out in vacuum, provided the magnitude, correlations, and decay of fluctuations about the average structure [25]. In 1988, Levitt and Sharon [26] calculated the 210-ps trajectory for the same protein, also including the surrounding water molecules. Since then, plenty of researches were carried out to understand protein dynamics by means of MD simulations, with a variety of applications [23, 24].

The computational burden for the calculation of the complete trajectories sometimes prevents the applicability of MD to large molecular complexes, especially when investigating the large-scale functional motions. For this reason, alternative strategies were developed in the following decades by scientists, in order to study protein dynamics in a more simplified and computationally efficient way. That is when Normal Mode Analysis (NMA) and Elastic Network Models (ENMs) came into play, being recognized as powerful tools to investigate protein dynamics and vibrations.

3.1   NORMAL MODE ANALYSIS AND THE ASSUMPTION OF THE HARMONIC POTENTIAL

In the early 1980s, it was discovered that the conformational energy surface of globular proteins near the energy minimum is approximately a multi-dimension parabola within the range of thermal fluctuations [27, 28]. This suggested the possibility of using a harmonic approximation for the exploration of the small-amplitude fluctuations, supporting the idea that proteins might behave like elastic bodies. This discovery opened up the way for the application of NMA to the field of protein and macromolecule dynamics [29].

The purpose of NMA is to describe the principal collective motions of a group of N atoms moving in a potential energy function V [30]. Empirical potential energies often used in MD simulations can be expressed in the following form:

$$V(r_1, r_2, \ldots, r_N) = \sum_{bonds} k_b (r - r_0)^2 + \sum_{angles} k_\theta (\theta - \theta_0)^2$$

$$+ \sum_{dihedrals} \frac{v_n}{2} [1 + \cos(n\phi - \gamma)] \qquad (3.1)$$

$$+ \frac{1}{2} \sum_{i,j} \left(\frac{q_i q_j}{\epsilon_r} + \frac{A}{r^{12}} - \frac{B}{r^6} \right),$$

where the first and second terms describe the harmonic potential involved into bond stretching and bending, the third term refers to variations of dihedral angles, and the last term is associated with the Coulomb's electrostatic potential and Van der Waals interactions. Instead of solving the equations of motion based on the complete potential V, NMA assumes that the atoms undergo small oscillations around the equilibrium position (r_0 and θ_0) at characteristic frequencies f and following specific displacement patterns δ [30]. The potential energy can then be developed in Taylor series about the equilibrium and truncated up to the second-order terms:

$$V(r_1, r_2, \ldots, r_N) \cong \frac{1}{2} \sum_{i,\alpha} \sum_{j,\beta} \left(\frac{\partial^2 V}{\partial r_{i,\alpha} \partial r_{j,\beta}} \right)_0 \Delta r_{i,\alpha} \Delta r_{j,\beta}, \qquad (3.2)$$

where α and β represent one of the Cartesian directions x, y, and z, i, and j represent two generic atoms, and $r_{i,\alpha}$ refers to the α-coordinate of the ith atom after deformation [30]. Equation (3.2) can be rewritten in compact form as:

$$V(r_1, r_2, \ldots, r_N) \cong \frac{1}{2} u^T H u, \tag{3.3}$$

where H is the $3N \times 3N$ Hessian matrix containing the second-order derivatives of the potential energy calculated in the equilibrium position from Equation (3.2), and u is the $3N \times 1$ vector containing the displacements of the N atoms [30]. Including the information about the kinetic energy T, which can be written as a function of the $3N \times 3N$ mass matrix M and the vector of linear velocities $\dot{u}$ as:

$$T(r_1, r_2, \ldots, r_N) \cong \frac{1}{2} \dot{u}^T M \dot{u}, \tag{3.4}$$

and looking for a harmonic solution of the displacements u in the form:

$$u = \delta e^{-i\omega t}, \tag{3.5}$$

and finally considering the Lagrangian of the problem along with the Euler's equations [30], one obtains the characteristic eigenvalue-eigenvector dynamic equation:

$$H\delta_n = \omega_n^2 M \delta_n. \tag{3.6}$$

The numerical solution of Equation (3.6) allows to obtain the complete set of $3N$ eigenvalues ω_n and eigenvectors δ_n. The former are associated with the vibrational frequencies $f_n (f_n = \omega_n/2\pi)$, while the latter represent the dynamic mode shapes.

In 1983, Brooks and Karplus [31] and Go and colleagues [32] applied NMA to study the normal modes and vibrational frequencies of BPTI. Brooks and Karplus [31] made use of all the energetic terms associated with bond lengths, bond angles, as well as dihedral angles, while Go and colleagues [32] treated bond lengths and bond angles as fixed and treated only dihedral angles of the main backbone and side chains as the dynamic variables. Based on the obtained normal modes, the root-mean-square (rms) fluctuations $\langle \Delta r_i^2 \rangle$ of the ith atom were calculated as:

$$\langle \Delta r_i^2 \rangle = k_B T \sum_n \frac{|\delta_{n,i}|^2}{\omega_n^2}, \tag{3.7}$$

$\delta_{n,i}$ being the vector of the nth normal mode with angular frequency ω_n for the ith atom, k_B is the Boltzmann constant and T is the absolute temperature.

The vibrational frequencies of BPTI obtained by Go and colleagues [32] and by Brooks and Karplus [31] are shown in Figures 3.1a and 3.1b, respectively. Notice that frequencies are expressed as the corresponding light wave numbers, where the wave number is defined as the frequency (in Hz) divided by the speed of light. Accordingly, the units of measurement of wave numbers turn out to be m^{-1}, and they are commonly expressed in cm^{-1} (1 cm^{-1} corresponding

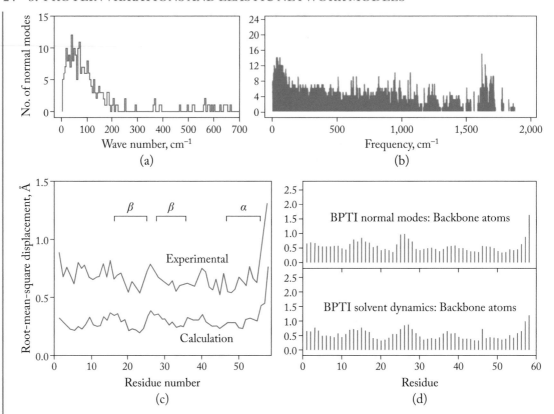

Figure 3.1: NMA of BPTI: vibrational frequencies obtained by (a) Go et al. [32] and (b) Brooks and Karplus [31]; rms fluctuations obtained by (c) Go et al. [32] and (d) Brooks and Karplus [31]. (a,c) Reprinted with permission from Go N. et al. Dynamics of a small globular protein in terms of low-frequency vibrational modes. *Proc. Natl. Acad. Sci.*, 80:3696–3700, 1983 [32]. (b,d) Reprinted with permission from Brooks B. and Karplus M. Harmonic dynamics of proteins: Normal modes and fluctuations in bovine pancreatic trypsin inhibitor. *Proc. Natl. Acad. Sci.*, 80:6571–6575, 1983 [31].

to about 30 GHz). By using only dihedral angles as the dynamic variables (Figure 3.1a), a total of 241 normal modes was found, 80% of which have frequencies below 200 cm^{-1} (6 THz), with a significant clustering peak in the spectrum around 50 cm^{-1} (1.5 THz). The first low-frequency mode was found to vibrate at 5.7 cm^{-1} (0.17 THz). When all the degrees of freedom (DOFs) were included into the calculation, a broader frequency spectrum was obtained (Figure 3.1b). Nevertheless, a similar pronounced peak around 50 cm^{-1} was still found and the first lowest-frequency mode was obtained at 3.1 cm^{-1} (0.09 GHz). The large clusters above 1,200 cm^{-1} (36 THz) were found to involve mainly bond-stretching vibrations.

Root-mean-square fluctuations of BPTI are shown in Figures 3.1c and 3.1d according to the work of Go [32] and Brooks [31], respectively. In Figure 3.1c, the calculated rms fluctuations from Equation (3.7), with small corrections due to anharmonic terms, are compared to the experimental values obtained by means of X-ray crystallography. Because corrections for contributions from static lattice disorder are not made in the upper experimental curve, quantitative comparison cannot be made between the calculated and experimental curves. However, it can be seen how the qualitative agreement of the general trend is good. This indicates the usefulness of the description of the dynamic structure in terms of collective variables corresponding to the normal modes of vibration [32]. Figure 3.1d shows the normal-mode rms fluctuations calculated at 300 K, compared with the results of an MD simulation of the same protein in a van der Waals solvent. As can be seen, the MD and normal mode trends are very similar, with slight differences mostly due to anharmonic and solvent effects [31].

Two years later, Levitt et al. [33] published a study where NMA in internal coordinates (torsion angles) was applied to four proteins: BPTI, crambin, ribonuclease, and lysozyme. The analysis started with the energy minimization of the X-ray crystal structure and the calculation of the second-order derivatives matrices related to the kinetic and potential energy of the system. The diagonalization of such matrices ultimately provided the vibrational frequencies and mode shapes associated with the harmonic motions. The investigation of the frequency spectra and the fluctuations calculated from the normal modes once again confirmed that NMA is able to properly describe the low-frequency protein dynamics. The mode-based fluctuations were compared to the B-factors arising from X-ray crystallography experiments, these ones being related to the rms fluctuations by the following equation:

$$B_i = \frac{8}{3}\pi^2 \langle \Delta r_i^2 \rangle, \tag{3.8}$$

providing correlation coefficients up to 0.86 [33].

Another important aspect captured by NMA is that the obtained low-frequency modes are usually found to be highly collective (nonlocal) and that these are the ones contributing for the most part to the atomic fluctuations. Figure 3.2 shows the lowest-frequency motions obtained for BPTI, ribonuclease and lysozyme, which were found to vibrate at 4.56 cm^{-1} (0.14 THz), 2.43 cm^{-1} (0.07 THz) and 2.98 cm^{-1} (0.09 THz), respectively [33]. As can be seen, these motions are highly collective, as they involve the concerted motion of most parts of the protein structure. As we will see in the remainder of this chapter and especially in the following one, the investigation of these low-frequency motions can reveal crucial insights regarding the biological mechanism of the protein under scrutiny.

Despite slight differences depending on the actual protein shape and flexibility, it is noteworthy that proteins were found to exhibit vibrations in the same regions of the frequency spectrum. In 1993, ben-Avraham compared the NMA frequency spectra of five different globular proteins (g-actin, lysozyme, ribonuclease I, BPTI, and crambin) and found that these spectra share many common features [34]. As shown in Figure 3.3a, all the curves associated with the

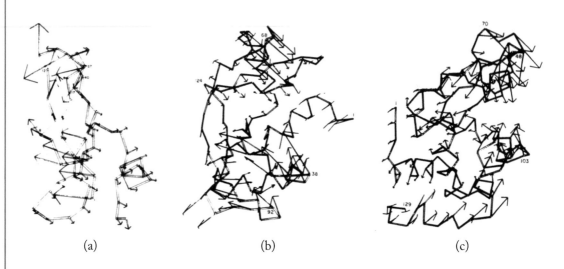

(a)   (b)   (c)

Figure 3.2: Lowest-frequency motions of (a) BPTI, (b) ribonuclease, and (c) lysozyme. Reprinted from Levitt et al. Protein normal-mode dynamics: Trypsin inhibitor, crambin, ribonuclease, and lysozyme. *J. Mol. Biol.*, 181:423–447, 1985 [33]. Copyright 2021 with permission from Elsevier.

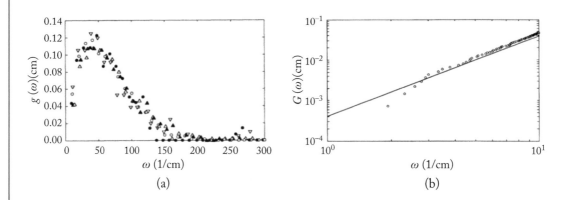

(a)   (b)

Figure 3.3: (a) Density of vibrational normal modes, $g(\omega)$, of g-actin, lysozyme, ribonuclease I, BPTI, and crambin as a function of frequency, (b) fraction of normal modes below frequency ω, $G(\omega)$, as a function of frequency for the slowest 70 modes of g-acting. Reprinted with permission from Ben-Avraham D. Vibrational normal-mode spectrum of globular proteins. *Phys. Rev. B*, 47:14559–14560 [34]. Copyright 2021 by the American Physical Society.

normalized density of the normal modes computed through NMA, $g(\omega)$, seem to collapse into a universal curve, which might then be representative of the frequency spectrum of all globular proteins.

Moreover, by analyzing the distribution in the lowest-frequency range, the author investigated the fraction of total normal modes up to frequency ω, i.e., $G(\omega)$, as a function of frequency. From the results, which are shown in logarithmic scale in Figure 3.3b, it was concluded that $G(\omega)$ scales as ω^2, and therefore $g(\omega)$ scales as ω [34]. This result suggested that, in the low-frequency range, protein might behave more like two-dimensional objects, rather than three-dimensional ones.

As shown in this section, NMA provides valuable information regarding the vibrational frequencies of proteins as well as their low-frequency motions. Nevertheless, two main drawbacks of classical NMA were: (i) the use of complex phenomenological potentials to model the covalent and nonbonded interactions between atom pairs, as reported in Equation (3.1) and (ii) the necessity of carrying out an initial energy minimization of the structure, which might be computational unfeasible for larger macro-molecular complexes and might also lead to unrealistic energetically-minimized structures. To overcome these criticalities, in 1996 Tirion opened up the way for the use of simplified elastic models to study protein dynamics and vibrations [35]. The full potential reported in Equation (3.1) was replaced with one of a much simpler form based on Hookean pairwise interactions:

$$V(r_1, r_2, \ldots, r_N) = \frac{C}{2} \sum_{i,j} \left(\left| r_{i,j} \right| - \left| r_{i,j} \right|_0 \right)^2, \tag{3.9}$$

$\left| r_{i,j} \right|$ and $\left| r_{i,j} \right|_0$ being, respectively, the distance between atoms i and j in the deformed and reference structure, respectively, and C representing the strength of the potential well, i.e., the stiffness, assumed to be equal for all interacting pairs. The sum reported in Equation (3.9) was actually restricted only to certain pairs of atoms, based on the criterion of geometrical proximity, i.e., only pairs of atoms whose spatial distance is less than a certain cutoff value r_c were considered. In the atomistic model proposed by Tirion, the value of r_c was fixed as $R_{vdW}(i) + R_{vdW}(j) + R_c$, $R_{vdW}(i)$ and $R_{vdW}(j)$ being the van der Waals radii of atoms i and j, respectively, and R_c an arbitrary cutoff parameter which models the decay of these elastic interactions [35].

Figure 3.4a shows the fraction of normal modes below a certain frequency ω, i.e., $G(\omega)$, for g-acting:ADP:Ca^{++} for the first 138 modes. The dashed curve refers to the data obtained using the complex L79 potential [36], while the continuous lines refer to the Tirion's method obtained with cutoff parameters R_c of 1.1, 1.5, 2.0, and 2.5 Å, with C values equal to 2.49, 1.29, 0.73, and 0.47 kJ/Å^{2}mol, respectively [35]. As can be seen, despite the simplicity of the single-parameter harmonic potential from Equation (3.9), the obtained low-frequency values are in line with the results from full NMA calculations. Figure 3.4b shows the comparison between the computed fluctuations (B-factors) obtained from the L79 potential (dashed line)

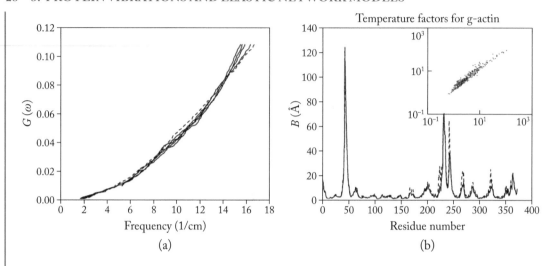

Figure 3.4: (a) Fraction of normal modes below frequency ω, $G(\omega)$, as a function of frequency for the slowest 150 modes of the g-acting:ADP:Ca^{++} system, (b) comparison of the theoretical B-factors obtained with the L79 potential (dashed curve) and the potential from Equation (3.9) (continuous curve). Reprinted with permission from Tirion M. M. Large-amplitude elastic motions in proteins from a single-parameter, atomic analysis. *Phys. Rev. Lett.*, 77:1905–1908, 1996 [35]. Copyright 2021 by the American Physical Society.

and the single-parameter Tirion's potential with $R_c = 2.2$ Å (continuous line). The inset shows the scatter plot of the two data set, highlighting the high agreement between the results [35].

Based on these results, Tirion was able to show that a simpler single-parameter harmonic potential was sufficient to reproduce the slow (low-frequency) protein dynamics in good detail. Moreover, the costly and sometimes inaccurate energy minimizations were eliminated, thus permitting the direct analyzing of the protein crystal structures available on databases [35]. Elastic models started then to be widely used by computational biologists and researchers to investigate protein dynamics and vibrations, the protein structure being modeled as a simple elastic network of Hookean springs among all the atoms.

A step toward further simplification of the protein model, in order to reduce the computational cost as well as to focus more on the low-frequency motions, was based on the adoption of coarse-grained strategies. Coarse-grained models are based on the idea that certain properties of a system with a large number of DOFs can be still retrieved with a simplification of the model by concentrating the complete DOFs into just a few of them [37]. Coarse-grained models for proteins usually involve the clustering of several atoms into one single bead in a representative location. Designing a particular coarse-grained model requires three main aspects: (i) the basic simulation unit, i.e., the number and location of these beads; (ii) the effective coarse-grained energy function, which should be able to simulate the interactions between the beads, usually based on considerations from the complete physical model; and (iii) the effective dynamical equations describing the time-evolution of the coarse-grained model [38]. Obviously, by adopting higher coarse-graining strategies for the protein structure, one is able to reduce the computational burden, at the expenses of a general decrease in accuracy and increase in difficulty for the definition of the coarse-grained energy functions. Coarse-grained models have been extensively used in the past decades to describe protein dynamics and functionality via MD simulations and NMA [37]–[40]. Putting together the Tirion's (all-atom) elastic model with the coarse-grained approach, the ENMs were then developed.

3.2   COARSE-GRAINED ELASTIC NETWORK MODELS

In 1997, Bahar et al. [41] and Haliloglu et al. [42] proposed using of a simplified single-parameter coarse-grained model for the investigation of protein dynamics and fluctuations. This new model was named Gaussian Network Model (GNM) as it was based on the assumption that interatomic distances follow a Gaussian distribution around their equilibrium values [41].

According to the GNM, the protein is modeled as a simple network of Hookean springs connecting representative nodes of the structure, located at the positions of the C^α atoms. The fluctuations $\Delta r_{i,j}$ in the separation distances $r_{i,j} = |r_j - r_i|$ between the ith and jth C^α atom in the native structure are assumed to obey the following Gaussian distribution:

$$W\left(\Delta r_{i,j}\right) = \left(\frac{\gamma}{\pi}\right)^{3/2} \exp\left(-\gamma \Delta r_{i,j}^2\right), \tag{3.10}$$

where γ is the counterpart of the single-parameter spring constant used by Tirion [35]. Following the statistical theory of polymer networks, the distribution function $W(\Delta r_{i,j})$ is substituted into the expression $\Delta A = -k_B T \ln W(\Delta r_{i,j})$ for the elastic free energy change associated with the fluctuation $\Delta r_{i,j}$ [41]. Then, the partition function for a protein made up of N residues can be expressed as:

$$Z_N = M \exp\left(\Delta r^{\mathrm{T}} \Gamma \Delta r\right), \tag{3.11}$$

where Δr is a N-dimensional vector composed of all the fluctuations Δr_i of each C^α atom, M is a constant, and Γ is a $N \times N$ symmetric matrix, usually referred to as the Kirchhoff or connectivity matrix [41]. This matrix contains the information about the connectivity among all the C^α atoms in the protein native structure. The elements of Γ, i.e., $\Gamma_{i,j}$, can be expressed as:

$$\Gamma_{i,j} = \begin{cases} -\gamma, & i \neq j, \ r_{i,j} \leq r_c \\ 0, & i \neq j, \ r_{i,j} > r_c \\ -\sum_{i \neq j} \Gamma_{i,j}, & i = j, \end{cases} \tag{3.12}$$

where r_c is the cutoff separation distance defining the range of nonbonded interacting contacts, i.e., C^α atoms closer than r_c are assumed to be interacting and then they are connected by a spring with stiffness γ, otherwise they are not considered interacting and no spring is generated between them. Based on considerations related to first interaction shells and to best fits with experimental data, r_c in GNM is usually taken as 7.0 Å [41, 42].

The cross-correlations between the fluctuations of two atoms k and l can be obtained by using the Boltzmann equation:

$$\langle \Delta r_k \cdot \Delta r_l \rangle = \frac{1}{Z_N} \int \Delta r_k \cdot \Delta r_l \exp\left(-\frac{\Delta r^{\mathrm{T}} \Gamma \Delta r}{k_B T}\right) d\Delta r = \left[\Gamma^{-1}\right]_{k,l}. \qquad (3.13)$$

The atomistic fluctuations of each ith C^α atom, i.e., $\langle \Delta r_i^2 \rangle$, can be derived from Equation (3.13) by simply putting $i = k = l$, therefore, they are equal to the diagonal i–i element of the inverse Kirchhoff matrix. By applying Equation (3.8), one can then simply evaluate the theoretical B-factor associated with each particular atom based on the GNM Γ^{-1} matrix. Off-diagonals elements of this matrix stand for the cross-correlations of the fluctuations between different residues.

It has to be observed that, since the determinant of Kirchhoff matrix is equal to zero, the matrix cannot be inverted directly. The inverse can then be calculated based on the modal decomposition, i.e.:

$$\Gamma = U\Lambda^2 U^{\mathrm{T}}, \qquad (3.14a)$$

$$\Gamma^{-1} = U\left(\Lambda^2\right)^{-1} U^{\mathrm{T}}, \qquad (3.14b)$$

where U is an orthogonal matrix whose columns correspond to the eigenvectors of Γ, and Λ^2 is a diagonal matrix whose entries λ^2 are the eigenvalues of the Kirchhoff matrix [41]. Due to the properties of the Kirchhoff matrix, the first eigenvalue λ_1^2 is equal to zero and the corresponding eigenvector represents an overall rigid translation of the protein structure. Therefore, the calculation of the inverse Kirchhoff matrix shown in Equation (3.14b) is carried out after eliminating the contribution of the zero eigenvalue and considering the remaining $N - 1$ modes.

As can be easily recognized, Equations (3.14) resemble the eigenvalue decomposition already shown in Equation (3.6) for NMA, but with some differences: (i) in the GNM case, only one-dimensional fluctuations are evaluated for each C^α atom; (ii) the model is based on a coarse-grained representation and on a simplified single-parameter harmonic potential; and (iii) there is no information related to the mass of the system, since the Kirchhoff matrix takes only into account the connectivity and stiffness distribution within the protein structure. Based on these considerations, it follows that the GNM eigenvalues λ_i^2 and eigenvectors U are related to the vibrational frequencies and mode shapes of the proteins, but only in a qualitative and one-dimensional perspective, respectively. Better said, from the GNM, we do have information about the flexibility and motion of the molecules, but we don't have the quantitative information about the frequency of vibration nor the directionality of the motion.

The GNM methodology was tested on several protein structures, in order to predict their flexibility and motions. For example, Bahar et al. [41] tested the GNM on 12 protein structures,

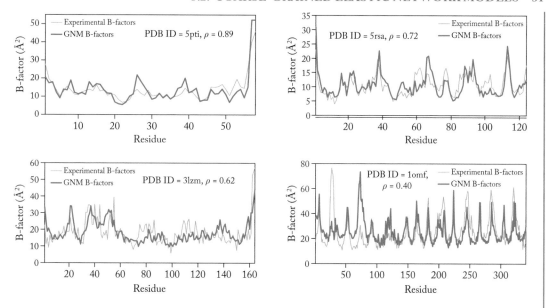

Figure 3.5: Comparison between experimental B-factors (thin lines) and GNM-based B-factors (thick lines) for various proteins of different sizes. The PDB ID of each case is reported within the panel, with the obtained correlation coefficient between numerical and experimental values. GNM calculations run from some of the examples reported in Bahar et al. [41].

and good results were found when comparing the theoretical with the experimental B-factors (Figure 3.5), as well as when comparing the theoretical cross-correlations from Equation (3.13) with those obtained by Levitt et al. [33] using NMA. Bahar and Jernigan [43] used the GNM for understanding the cooperative fluctuations and subunit communications in tryptophan synthase. Bahar et al. [44] applied the GNM to the large heterodimeric retroviral reverse transcriptase (HIV-1 RT), showing that the method is indeed able to provide good results even when applied to large complexes.

In the following years, various modifications were proposed for the GNM, with the goal of improving the accuracy of the numerical predictions. Micheletti et al. [45] introduced a different coarse-graining strategy, considering two beads for each amino acids, namely both the C^α atoms for the backbone and the C^β atoms for the sidechains. Yang et al. [46] developed a parameter-free GNM (pfGNM), where the necessity for the arbitrary cutoff parameter r_c was eliminated and the spring constants γ varied depending on the inter-residue distances in the protein structure with an inverse power dependence, i.e., $\gamma_{i,j} \sim r_{i,j}^{-2}$. Zhang and Kurgan [47] included the information about the amino acid sequence into the model, while more recently Zhang et al. [48] took also into account the relative solvent accessibility (RSA).

Despite the high simplification of the model, the GNM was shown to be able to properly reproduce protein dynamics and its flexibility. This suggests that these features are mostly dependent on the overall protein shape and the connections among the residues in the native structure. Therefore, using a simplified harmonic potential coupled with a coarse-grained protein representation is indeed able to reproduce the protein dynamics in good detail with low computational effort. However, the main shortcoming of the GNM is that it only provides information about the amplitudes of motion, but not its directions in the three-dimensional space. To overcome this limitation, Atilgan et al. [49] developed the Anisotropic Network Model (ANM), which starts from the same assumptions of the GNM (coarse-grained model with beads located at C^α atom positions and single-parameter harmonic potential), but it also considers the anisotropy of the protein motion in the three spatial directions.

The ANM considers a three-dimensional network of elastic springs as a representative model based on which to evaluate the protein fluctuation dynamics. As in the GNM, springs only connect those C^α atoms whose distance is lower than a pre-selected cutoff parameter r_c. If one considers two connected residues i and j, similar to Equation (3.9), the elastic potential stored within the connecting spring can be written as:

$$V_{i,j} = \frac{1}{2}\gamma \left(\left| \boldsymbol{r_{i,j}} \right| - \left| \boldsymbol{r^0_{i,j}} \right| \right)^2, \tag{3.15}$$

where γ represents the spring constant, while $\boldsymbol{r_{i,j}}$ and $\boldsymbol{r^0_{i,j}}$ are the inter-residue distances in the current and initial conformation of the network (Figure 3.6). The second derivatives of the potential $V_{i,j}$ with respect to the components of $\boldsymbol{r_i}$ (x_i, y_i, z_i) and $\boldsymbol{r_j}$ (x_j, y_j, z_j) are [49]:

$$\frac{\partial^2 V_{i,j}}{\partial x_i{}^2} = \frac{\partial^2 V_{i,j}}{\partial x_j{}^2} = \gamma \left(1 + \frac{\left| \boldsymbol{r^0_{i,j}} \right| (x_j - x_i)^2}{\left| \boldsymbol{r_{i,j}} \right|^3} - \frac{\left| \boldsymbol{r^0_{i,j}} \right|}{\left| \boldsymbol{r_{i,j}} \right|} \right), \tag{3.16a}$$

$$\frac{\partial^2 V_{i,j}}{\partial y_i{}^2} = \frac{\partial^2 V_{i,j}}{\partial y_j{}^2} = \gamma \left(1 + \frac{\left| \boldsymbol{r^0_{i,j}} \right| (y_j - y_i)^2}{\left| \boldsymbol{r_{i,j}} \right|^3} - \frac{\left| \boldsymbol{r^0_{i,j}} \right|}{\left| \boldsymbol{r_{i,j}} \right|} \right), \tag{3.16b}$$

$$\frac{\partial^2 V_{i,j}}{\partial z_i{}^2} = \frac{\partial^2 V_{i,j}}{\partial z_i{}^2} = \gamma \left(1 + \frac{\left| \boldsymbol{r^0_{i,j}} \right| (z_j - z_i)^2}{\left| \boldsymbol{r_{i,j}} \right|^3} - \frac{\left| \boldsymbol{r^0_{i,j}} \right|}{\left| \boldsymbol{r_{i,j}} \right|} \right). \tag{3.16c}$$

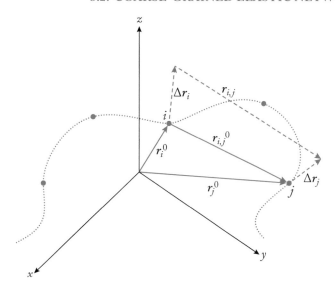

Figure 3.6: Schematic representation of the fluctuation vectors $\boldsymbol{\Delta r_i}$ and $\boldsymbol{\Delta r_j}$ in correspondence of the ith and jth residue. Vectors $\boldsymbol{r_i^0}$ and $\boldsymbol{r_j^0}$ refer to the initial positions of the residues in the native structure, whereas $\boldsymbol{r_{i,j}^0}$ and $\boldsymbol{r_{i,j}}$ denote their distance in the native and deformed configuration, respectively.

Evaluating such derivates at the equilibrium position, i.e., for $\boldsymbol{r_{i,j}}$ and $\boldsymbol{r_{i,j}^0}$, it follows:

$$\frac{\partial^2 V_{i,j}}{\partial x_i^2} = \frac{\partial^2 V_{i,j}}{\partial x_j^2} = \gamma \frac{(x_j - x_i)^2}{\left| \boldsymbol{r_{i,j}} \right|^2}, \tag{3.17a}$$

$$\frac{\partial^2 V_{i,j}}{\partial y_i^2} = \frac{\partial^2 V_{i,j}}{\partial y_j^2} = \gamma \frac{(y_j - y_i)^2}{\left| \boldsymbol{r_{i,j}} \right|^2}, \tag{3.17b}$$

$$\frac{\partial^2 V_{i,j}}{\partial z_i^2} = \frac{\partial^2 V_{i,j}}{\partial z_i^2} = \gamma \frac{(z_j - z_i)^2}{\left| \boldsymbol{r_{i,j}} \right|^2}. \tag{3.17c}$$

Similarly, the second cross-derivatives become:

$$\frac{\partial^2 V_{i,j}}{\partial x_i \partial y_j} = \frac{\partial^2 V_{i,j}}{\partial x_j \partial y_i} = -\gamma \frac{(x_j - x_i)(y_j - y_i)}{|r_{i,j}|^2}, \tag{3.18a}$$

$$\frac{\partial^2 V_{i,j}}{\partial x_i \partial z_j} = \frac{\partial^2 V_{i,j}}{\partial x_j \partial z_i} = -\gamma \frac{(x_j - x_i)(z_j - z_i)}{|r_{i,j}|^2}, \tag{3.18b}$$

$$\frac{\partial^2 V_{i,j}}{\partial y_i \partial z_j} = \frac{\partial^2 V_{i,j}}{\partial y_j \partial z_i} = -\gamma \frac{(y_j - y_i)(z_j - z_i)}{|r_{i,j}|^2}. \tag{3.18c}$$

Based on the computed second derivatives of the elastic potentials associated with all the Hookean springs connecting each couple of residues i and j, the ANM yields the calculation of the $3N \times 3N$ Hessian matrix H, which can be partitioned according to the N protein residues as [49]:

$$H = \begin{bmatrix} H_{1,1} & H_{1,2} & \cdots & H_{1,N} \\ H_{2,1} & H_{2,2} & \cdots & H_{2,N} \\ \cdots & \cdots & \cdots & \cdots \\ H_{N,1} & H_{N,2} & \cdots & H_{N,N} \end{bmatrix}, \tag{3.19}$$

where each sub-matrix $H_{i,j}$, with $i \neq j$, is the 3×3 Hessian matrix related to the spring connecting nodes i and j. It contains the second derivatives of the potential $V_{i,j}$ reported in Equations (3.17) and (3.18) and can then be expressed as:

$$H_{i,j} = \begin{bmatrix} \dfrac{\partial^2 V_{i,j}}{\partial x_i \partial x_j} & \dfrac{\partial^2 V_{i,j}}{\partial x_i \partial y_j} & \dfrac{\partial^2 V_{i,j}}{\partial x_i \partial z_j} \\ \dfrac{\partial^2 V_{i,j}}{\partial y_i \partial x_j} & \dfrac{\partial^2 V_{i,j}}{\partial y_i \partial y_j} & \dfrac{\partial^2 V_{i,j}}{\partial y_i \partial z_j} \\ \dfrac{\partial^2 V_{i,j}}{\partial z_i \partial x_j} & \dfrac{\partial^2 V_{i,j}}{\partial z_i \partial y_j} & \dfrac{\partial^2 V_{i,j}}{\partial z_i \partial z_j} \end{bmatrix}. \tag{3.20}$$

The diagonal sub-matrices of the Hessian matrix, i.e., $H_{i,i}$, can be obtained by summing up all the contributions of the jth nodes connected to the ith node, i.e.,

$$H_{i,i} = - \sum_{j=1, j \neq i}^{N} H_{i,j}. \tag{3.21}$$

Once the Hessian matrix is built from Equations (3.19), (3.20), and (3.21), it undergoes the eigenvalue-eigenvector decomposition. As in the case of the GNM Kirchhoff matrix, the determinant of H is zero, reflecting the lack of external constraint conditions to the protein network. In this case, we find six zero eigenvalues whose corresponding eigenvectors are associated with the six translation and rotation motions of the protein structure as a whole rigid body. The remaining $3N - 6$ non-zero eigenvalues λ_n are associated with the vibrational frequencies, and the corresponding eigenvectors δ_n to the non-rigid mode shapes of the protein structure. Notice that, similar to the GNM, the ANM does not take into account the protein mass, therefore the obtained eigenvalues are only qualitatively correlated to the vibrational frequencies.

The eigenvalues and eigenvectors of the ANM Hessian matrix can be used to calculate its pseudo-inverse, by using the same formal expression used for GNM in Equation (3.14)b, considering the $3N - 6$ non-zero eigenvalues and non-rigid eigenvectors. The pseudo-inverse $\boldsymbol{H}^{-1}$ can be seen as an $N \times N$ matrix made up of $3 \times 3 \boldsymbol{H}_{i,j}^{-1}$ blocks, whose elements constitute the cross-correlations between the fluctuation vectors associated with residues i and j. The diagonal blocks, i.e., $\boldsymbol{H}_{i,i}^{-1}$, are then used to compute the direct fluctuations of each residue i and its theoretical B-factor value through Equation (3.8). Note that, as in the GNM, the original version of the ANM assumes that all the springs have the same strength constant γ. Its absolute value can then be set up based on the comparison between the theoretical and experimental B-factors.

The ANM was applied by Atilgan et al. [49] to study the fluctuation dynamics and flexibility of the 183-residue retinol-binding protein (RBP). The values that have to be adopted for the spring constant γ as a function of the selected cutoff value r_c were investigated. It was found that, the higher the cutoff value, the lower the value of the spring constant. This mainly arises from the fact that an increase in the cutoff leads to a higher number of connections, therefore the value of the spring constant has to decrease in order to maintain the same level of flexibility. Since typical force constants are of the order of 1 kcal/Å^2, the researchers found that reasonable values of r_c for the ANM are 12–15 Å, respectively. Figure 3.7 shows the comparison between the experimental B-factors and the theoretical ones obtained from the ANM of RBP with a cutoff value of 13 Å. As can be seen, despite the simplification, the method is able to correctly identify the fluctuation dynamics of the protein.

The main advantage of the ANM over the GNM is that the former also enables to visualize the protein motions, as these are calculated with respect to the three-dimensional reference frame. As a result, the visualization of the ANM-based mode shapes offers the possibility to analyze the potential biological mechanisms of the protein structure under scrutiny. Figure 3.8a shows the motion associated with the second low-frequency motion of RBP, as obtained from Atilgan et al. [49], illustrating the large amount of flexibility of the loops near residues 65 and 95. Figure 3.8b shows the same protein (depicted in green), when it is complexed with its carrier protein transthyretin (in purple). As shown by the black arrows, the most flexible parts identified by the ANM motion are involved in the recognition of the transthyretin. This shows that ANM-based results have the potential to unravel the biological mechanisms of protein structures and macromolecular complexes.

In the subsequent years, the ANM was modified in order to obtain better correlations with the experimental data, i.e., the B-factors, by changing some fundamental model parameters. For example, Eyal et al. [50] evaluated the performance of the traditional ANM, in terms

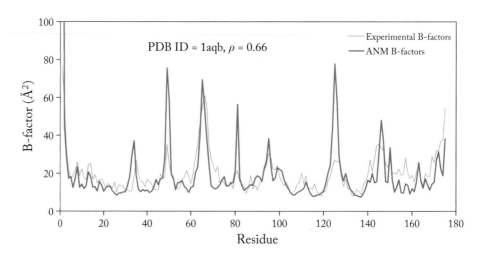

Figure 3.7: Comparison of experimental and ANM-derived B-factors for RBP obtained with $r_c = 13$ Å. ANM calculations run from the example reported in Atilgan et al. [49].

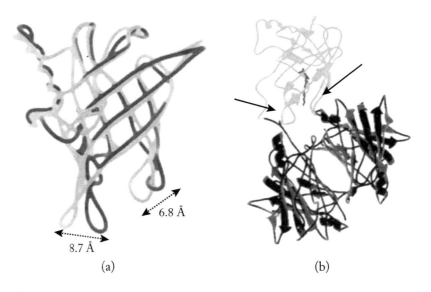

Figure 3.8: (a) Flexibility of RBP obtained by the second ANM mode, (b) complexation of RBP (in green) with its carrier protein transthyretin (in purple). Reprinted from Atilgan et al. Anisotropy of fluctuation dynamics of proteins with an elastic network model. *Bioph. J.*, 80:505–515, 2001 [49]. Copyright 2021 with permission from Elsevier.

of correlation with the experimental B-factors, for a set of 176 different protein structures. The influence of the cutoff distance r_c, in the range 10–24 Å, on the model performance was evaluated. From the results, the researchers found that a cutoff value around 18 Å is the one providing the high correlation with the experimental data. In the same paper, Eyal et al. [50] also proposed to consider distance weights for the spring constant, i.e., $\gamma_{i,j} \sim r_{i,j}^{-p}$, where the stiffness of each spring is inversely proportional to its length through the exponent p. The higher the value of p, the higher the dependence on the distance, while p values equal to zero reflect the ANM developed originally by Atilgan et al. [49], i.e., $\gamma_{i,j} = \gamma$ for each pair of residues. They found that by increasing this exponent in the range 0–2.8 the model leads to better agreements with the experimental B-factors. This might be due to the fact that in this way the distance-based decay of residue interactions within the protein network are represented more realistically.

A similar concept was applied by Yang et al. [46] a few years later while developing the parameter-free ANM (pfANM). In this model, which is similar to the pfGNM shown above, the authors completely remove the need for a geometrical cutoff distance r_c, as all the residues are connected by a spring with distance-dependent force constants $\gamma_{i,j} \sim r_{i,j}^{-2}$. The performance of both the ANM and pfANM was assessed for a set of 1,220 protein structures in terms of the correlation with the B-factors. It was found that including distance-dependent force constants leads to higher values of the correlation, therefore to an overall increase performance of the model [46].

As remarked above, both the GNM and ANM are able to describe the fluctuation dynamics of the protein structure around its equilibrium position based on the computation of the Kirchhoff and Hessian matrix, respectively. However, since the mass of the protein does not enter into the calculations in these models, the obtained eigenvalues are connected to the vibrational frequencies only in a qualitative manner. Therefore, the inclusion of mass into the model is needed to obtain information about the range of frequencies at which proteins vibrate. As shown in Section 3.1, NMA provides such frequency values, as it takes into account the mass of the system, although it uses more complex potentials than the ones employed in ENMs. In the following section, recent Structural Mechanics-based elastic models will be reported which allow to obtain a more quantitative understanding of the vibrational frequencies of proteins around their equilibrium position, but still maintaining the simplifying assumptions behind the ENMs.

3.3   FINITE ELEMENT ELASTIC LATTICE MODELS

Simplified calculations based on the wave propagation theory can provide a simplified estimate of the order of magnitude of the characteristic vibrations of proteins. Expansion-contraction vibrations generally induce longitudinal pressure waves traveling at a speed v which is characteristic of the medium (for most of solids and fluids, v being of the order of 10^3 m/s). The wavelength ψ cannot exceed the maximum size of the body involved into the vibration. Therefore, knowing the traveling speed v and the dimensional scale connected to ψ, the frequency of the pressure wave f can be obtained as $f = v/\psi$ [51]. For proteins, assuming a traveling speed of the pressure waves of about 10^3 m/s, we obtain vibrations at: (i) 10^{13} Hz for localized oscillations of chemical groups or amino acids ($\psi \approx$ Å $= 10^{-10}$ m); (ii) 10^{12} Hz for delocalized oscillations involving small proteins or extended portions of large proteins ($\psi \approx$ nm $= 10^{-9}$ m); and (iii) 10^{11} Hz for global oscillations involving large proteins ($\psi \approx 10^{-8}$ m) [51].

These orders of magnitude for the frequencies can also be obtained by the theory of mechanical vibrations of elastic systems. In this case, considering an elementary harmonic oscillator, made up of two atoms of mass m connected by a linear elastic spring of stiffness k, the fundamental frequency of vibration of the system is found to be $f = (k/m)^{1/2}/(2\pi)$. Using the values of the atom masses (of the order of 10^{-26} kg) and reasonable values of stiffness for inter-atomic covalent bonds (of the order of 200 N/m), one obtains a frequency of the order of 10^{13} Hz, which was already found for the vibrations of small chemical groups from the wave propagation theory [51].

Based on these considerations, Carpinteri et al. [51, 52] developed a frame-like finite element (FE) model where the expansions-contractions of the covalent bonds of the protein backbone were the basis to study the expansion-contraction vibrational modes of proteins. The model only considered the short-range covalent interactions among all the protein atoms and used the actual mass values for the atoms and average stiffness values for the covalent interactions. The latter were modeled as beam finite elements, with an axial stiffness $k \sim L^{-1}$ (L being the length of the element) and very high values for the bending, shear and torsional rigidity in order not to account for these modes of deformation [51, 52]. Modal analysis was then carried out based on the computed stiffness and mass matrices, i.e., K and M, numerically solving Equation (3.6). Note that here the Hessian matrix H has been replaced by the stiffness matrix K, which is the more common term used in the mechanics of solids and structures. The mass matrix M has been conveniently included into the model for the purpose of obtaining quantitative information about the frequencies of vibration. Although the method was not meant to provide a full description of the protein flexibility, it was able to provide good insights on the global expansion-contraction vibrations of hen egg-white lysozyme (Figure 3.9a) [51] and Na/K-ATPase [52], found to occur at THz and sub-THz frequencies (as obtained above from simplified hand calculations and confirmed by more complex NMA simulations).

The model proposed by Carpinteri et al. [51, 52] was then taken as the basis to develop a new FE elastic truss model which can be seen as the counterpart of ANM by following a purely Structural Mechanics approach. Namely, Scaramozzino et al. [53] and Giordani et al. [54] proposed a FE truss model, where the elastic connections are simply hinged at the nodes of the protein structure (Figure 3.9b). In [53] it was shown that this structural model leads to a stiffness matrix K, obtained by the FE approach as:

$$K = \sum_{i,j|r_c}^{N} C_{i,j}^{T} N_{i,j}^{T} k_{i,j}^{*} N_{i,j} C_{i,j}, \tag{3.22}$$

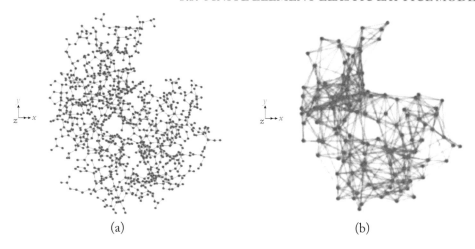

(a)   (b)

Figure 3.9: FE lattice models of hen egg-white lysozyme: (a) all-atom frame-like model and (b) coarse-grained truss model. (a) Reprinted from Carpinteri A. et al. Terahertz mechanical vibrations in lysozyme: Raman spectroscopy vs. modal analysis. *J. Mol. Struct.*, 1139:222–230, 2017, with permission from Elsevier [51]. (b) Reprinted by permission from Springer from Scaramozzino D. et al. A finite-element-based coarse-grained model for global protein vibration. *Meccanica*, 54:1927–1940, 2019 [53].

which is perfectly consistent with the ANM Hessian matrix H reported in Equation (3.19). In Equation (3.22), $k_{i,j}^*$, $N_{i,j}$, and $C_{i,j}$ are, respectively, the 2×2 stiffness matrix of the elastic element connecting nodes i and j expressed in its local reference system, the 2×6 rotation matrix which contains the direction cosines between the local reference system of the element and global reference system XYZ, the $6 \times 3N$ expansion matrix [1, 53].

The main difference with the ANM is that this FE truss model also considers the mass of the system, through the mass matrix M, thus allowing to obtain quantitative information about the vibrational frequencies, once the stiffness values of the elastic connections are designed based on the comparison with the experimental B-factors [53, 54]. Depending on the adopted cutoff values, the lowest vibrational frequencies were found to occur in the range 0.05–0.1 THz (1.5–4 cm^{-1}) [53], in agreement with the previous NMA studies from Levitt et al. on lysozyme [33]. The low-frequency mode shapes were also found to be in agreement with the previous Levitt's studies (see Figure 3.2c and 3.10a). From the analysis of the modal displacements of the truss structure, it was then possible to investigate the most flexible regions of the protein (Figure 3.10b), which ultimately reflect the experimental B-factors [53]. Once again, these results confirm that simplified elastic models are able to capture the overall vibrational behavior of proteins and their inherent flexibility features.

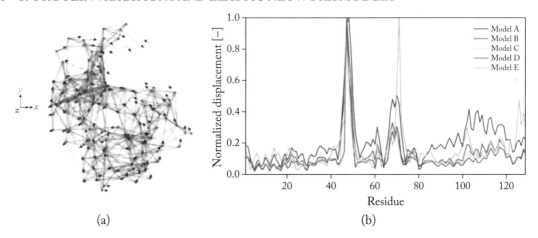

Figure 3.10: Lowest-frequency mode of hen egg-white lysozyme detected by the simplified FE truss model: (a) mode shape and (b) profile of normalized modal displacements. In (a), the mode shape of the model with $r_c = 8$ Å is shown, while in (b) the five colored curves refer to model A, B, C, D, and E, with r_c values equal to 8, 10, 12, 15, and 20 Å, respectively. Reprinted by permission from Springer from Scaramozzino D. et al. A finite-element-based coarse-grained model for global protein vibration. *Meccanica*, 54:1927–1940, 2019 [53].

3.4 FURTHER ELASTIC MODELS FOR PROTEIN VIBRATIONS

Other than the ENMs and FE-based elastic lattice models presented in the previous two sections, additional elastic models were developed during the last decades to describe protein dynamics and functional motions. At the end of the 20th century, Hinsen and colleagues made use of NMA and the harmonic assumption to capture the low-frequency vibrations of proteins [55]–[57]. In these works, the harmonic spring constants between the ith and jth nodes were assigned the following distance-dependence form:

$$\gamma_{i,j} = C \exp\left(-\frac{\left|r^0_{i,j}\right|^2}{r_c^2}\right), \tag{3.23}$$

being C a constant value, $|r^0_{i,j}|$ the distance between the nodes in the reference structure, and r_c a cutoff parameter, usually set as 3 Å and 7 Å for the all-atom and coarse-grained protein model, respectively. Moreover, Hinsen was among the first researchers to put great attention into the definition of rigid domains within the protein structure and look at the protein motion as overall translations and rotations of domains [55, 56].

This approach was developed further by Tama et al. [58] who developed the so-called RTB (rotations-translations of blocks). In this model, the protein is not described as a network of flexible springs connecting atoms or residues, but it is seen as a union of rigid blocks, whose translations and rotations drive its low-frequency motions. These blocks often contain several atoms or residues, which are supposed to move together as rigid bodies. Starting from the $3N \times 3N$ Hessian matrix of the system H, the RTB expresses this matrix in a smaller basis defined by the translations and rotations of n_b blocks, thus defining a projected Hessian matrix H_b, obtained as:

$$H_b = P^T H P, \tag{3.24}$$

where P is an orthogonal $3N \times 6n_b$ matrix built considering the vectors associated with the local translations and rotations of the blocks [58]. As a result, H_b turns out to be a $6n_b \times 6n_b$ matrix, from the diagonalization of which one obtains the low-frequency mode shapes of the protein, defined upon the translations and rotations of the unit blocks. Once the $6n_b \times 6n_b$ eigenvector matrix A_b is obtained from the diagonalization of H_b, the $3N \times 3N$ eigenvector matrix A related to the $3N$ atomic displacements can be retrieved by the simple expression $A = PA_b$. Finally, once the mode shapes are computed in the common basis of atomic displacements, these can be used to calculate the fluctuations by using Equation (3.7) as well as to visualize the motions of the protein structure in the three-dimensional space.

The method was applied by Tama et al. [58] to calculate the low-frequency vibrations of twelve proteins with different sizes, ranging from 46 (crambin) up to 858 residues (dimeric citrate synthase). From the results, RTB was shown to provide correct estimates of the protein fluctuations and to describe fairly accurately the low-frequency motions, even when blocks included up to six amino acids. It is clear that this approach is very efficient in terms of decreasing the computational burden. The higher the number of atoms/residues moving together as a rigid block, the lower the number of DOFs of the system, and therefore the lower the computational cost in order to obtain the information about the protein slow motions.

In 2017, Hoffmann and Grudinin [59] developed another model, called NOLB (NOn-Linear rigid Block), where the RTB modes are extrapolated in order to follow the curvilinear pathways involved by the block rotations. In this way, a more realistic description of the large-scale protein motions can be obtained, while avoiding the unrealistic deformations involved by the simplistic linear extrapolation of the RTB modes. More recently, we and the Jernigan's group developed a more comprehensive elastic model, which was called hdANM (hinge-domain Anisotropic Network Model) [60]. In the hdANM, the protein is modeled as a structure made up of both flexible and rigid parts, the former being associated with the hinges and the latter with the domains. The hdANM was proven to be suitable to describe the large-scale functional motions of proteins, while drastically reducing the computational cost. However, due to the fact that both NOLB and hdANM were purposely developed to describe the large-scale protein motions, rather than their small-amplitude fluctuations, we shall describe them in more detail in the next chapter, where we will see how these elastic modes are strongly connected with the protein biological mechanisms through the concept of conformational change.

CHAPTER 4

Protein Vibrations and Conformational Changes

We mentioned in Chapter 2 that proteins are not static entities, but flexible macromolecules able to change their three-dimensional conformation based on external conditions (alteration of the pH or temperature, presence of ligands, application of forces, etc.). These changes in the three-dimensional structure are usually known as the conformational changes of the protein and are pivotal for a variety of biological tasks. Kinesin undergoes a large conformational change, which allows the protein to transport its cargo by walking on microtubules. Hemoglobin experiences a quaternary conformational change when biding to oxygen molecules. Conformational changes are also observed in transmembrane proteins when they act as ion channels, letting ions pass through the cell membrane.

In Chapter 3, we saw that protein flexibility is closely associated with the internal protein dynamics. Since conformational changes require structural rearrangements of the three-dimensional structure, it is quite clear that protein flexibility has the potential to play a significant role in driving the conformational transition. It should not surprise then if protein vibrations (which, as we saw in the previous chapter, are related to protein flexibility) will be found to be associated with the conformational change.

4.1 PROTEIN CONFORMATIONAL CHANGES

At the end of Chapter 2 we introduced the fundamental paradigm of protein action, in the form sequence-structure-dynamics-function. This paradigm tells us that protein dynamical motions are the way through which a certain protein structure carries out its biological function. Protein motions can also be interpreted as those movements that lead the starting protein configuration to a certain final conformation, i.e., protein motions imply conformational changes.

Conformational changes in proteins can be small or large. In the former case, atoms usually undergo displacements far only few Å (or even less) from their reference structure. In the latter, they can even experience displacements of few nanometers. On the other hand, conformational changes can be localized or collective. In the former case, the conformational change only affects a small portion of the protein. In the latter, many regions of the protein are found to move collectively in order to complete the structural rearrangement. Figure 4.1 shows some examples of known conformational changes, where proteins exhibit different conformations depending on external factors. Figure 4.1a shows the "open" conformation of calmodulin, whereas Figure 4.1b

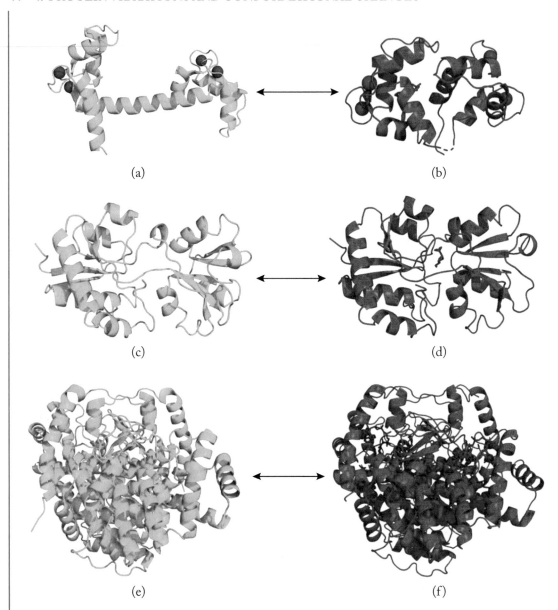

Figure 4.1: Examples of protein conformational changes: (a) open (PDB: 1cll) and (b) closed (PDB: 1ctr) conformation of calmodulin upon binding to trifluoperazine; (c) open (PDB: 2lao) and (d) closed (PDB: 1lst) conformation of LAO-binding protein upon binding to lysine; (e) open (PDB: 5csc) and (f) closed (PDB: 6csc) conformation of citrate synthase upon binding to diethyl phosphonate.

the "closed" form that the protein achieves when it binds to trifluoperazine. The double-arrow between the two structures indicates that there exists some motion leading calmodulin to switch from the open to the closed conformation, and vice versa. Figures 4.1c and 4.1d show the open and closed forms of LAO-binding protein upon binding to a molecule of lysine. Finally, Figures 4.1e and 4.1f report the conformational change of citrate synthase due to the binding-unbinding with diethyl phosphonate.

As said above, conformational changes can be large or small, collective, or localized. In computational biology, two simple numerical metrics are used, which provide estimates of the largeness and collectiveness of the conformational change. The former is often measured by the root-mean-square deviation (RMSD) of the atomic positions, while the latter by means of the degree of collectivity κ. Before reporting the mathematical expressions of the RMSD and κ, we shall briefly discuss the superimposition between the two protein conformations.

Imagine that you know from experimental tests, e.g., X-ray crystallography, two conformations of the same protein. These conformations can usually be obtained from the files available in public databases, such as the PDB [21]. Besides plenty of other information, the PDB files contain all the coordinates of the protein structure in a certain reference frame XYZ. When comparing the two conformations A and B of a certain protein, you need to consider the fact that the reference frames adopted for the resolution of these structures are different. Therefore, you cannot evaluate the displacements associated with the conformational change simply by taking the difference of these coordinates. Before doing that, the two protein conformations must be superimposed, i.e., their coordinates must be expressed in the same reference system.

To do so, let's take A as the reference protein conformation and B as the target conformation. The difference in the reference systems XYZ related to A and B can be expressed by means of a rigid translation and rotation of these frames. We can directly assign this rigid translation and rotation to the reference structure A, so that the coordinates of B can be written as:

$$B = RA + T + C.\tag{4.1}$$

In Equation (4.1), B and A are the $3N \times 1$ vectors containing the XYZ coordinates of the N atoms of the conformation B and A, respectively, T and R are the $3N \times 1$ and $3N \times 3N$ matrices accounting for the rigid translation and rotation, respectively, and C is the $3N \times 1$ vector which takes in account only the internal protein deformation. That is, C corresponds to the actual displacements of the conformational change. Known the coordinate vectors A and B and posing that the norm of the vector C is minimum, i.e., applying the least-squares method, one finds the estimates of the translation and rotation matrices, T' and R'. The conformation B can then be expressed in the same reference system of A, i.e.:

$$B_{TR} = R'^{-1}(B - T'),\tag{4.2}$$

where B_{TR} is the vector containing the roto-translated coordinates of B in the reference system of A. Equation (4.2) reflects the superimposition of B over A. At this point, the vector C

reflecting the displacements of the conformational change can be simply evaluated as:

$$C = B_{TR} - A = R'^{-1}(B - T') - A. \tag{4.3}$$

C contains the displacements along the XYZ axes of the A–to–B conformational change, expressed in the reference system of the reference conformation A. Based on the values of the displacements contained in C, we can finally investigate the largeness and the collectivity of the conformational change, as mentioned above. The former can be assessed by means of the RMSD, which is defined as:

$$RMSD = \sqrt{\frac{1}{N}|C|^2} = \sqrt{\frac{1}{N}\sum_{i=1}^{N}\left(c_{i,x}^2 + c_{i,y}^2 + c_{i,z}^2\right)}, \tag{4.4}$$

where $|C|$ stands for the norm of vector C, N is the total number of atoms used to calculate the RMSD (usually N is equal to all the heavy atoms of the protein for all-atom representations, or only the C^α atoms when dealing with coarse-grained approaches), and $c_{i,x}$, $c_{i,y}$, and $c_{i,z}$ represent the X-, Y-, and Z-component of the displacement of the ith atom. For instance, the conformational change of calmodulin (Figures 4.1a and 4.1b) exhibits an RMSD, calculated over all the heavy atoms of the protein, equal to 12.3 Å. Conversely, LAO-binding protein and citrate synthase show RMSDs of 4.5 Å and 1.4 Å, respectively. These values inform us that calmodulin has a relatively large-scale conformational change, LAO-binding a medium-scale one, while citrate synthase has only a slight conformational rearrangement (Figure 4.1).

On the other hand, each conformational change can be collective or localized. The degree of collectivity κ tells us how much a certain motion is collective. This parameter is definied as follows [61]:

$$\kappa = \frac{1}{N}\exp\left[-\sum_{i=1}^{N}\left(\frac{c_i^2}{\sum_{i=1}^{N}c_i^2}\right)\ln\left(\frac{c_i^2}{\sum_{i=1}^{N}c_i^2}\right)\right], \tag{4.5}$$

where $c_i^2 = c_{i,x}^2 + c_{i,y}^2 + c_{i,z}^2$. As in the case of the RMSD, the summations in Equation (4.5) can be carried out considering all the heavy atoms of the protein or only the C^α atoms, depending on the adopted protein representation. From a careful observation of Equation (4.5), it follows that, if the motion is completely collective, i.e., all the atoms move together of the same quantity $c_i = c$, the degree of collectivity κ is equal to 1. On the other hand, when the motion is restricted to one single atom, κ reaches it minimum value of $1/N$ (which is a very small number, almost close to zero). Therefore, values of κ close to 0 represent highly localized motions, whereas values close to 1 describe highly collective motions. For the cases reported in Figure 4.1, we find $\kappa = 0.68$, 0.68, and 0.30 for calmodulin, LAO-binding protein and citrate synthase, respectively [61]. These values indicate that the conformational changes of calmodulin and LAO-binding protein are highly collective, whereas that of citrate synthase is rather confined to only few portions of the protein structure (Figure 4.1).

4.2 PROTEIN LOW-FREQUENCY VIBRATIONS PREDICT THE DIRECTION OF THE CONFORMATIONAL CHANGE

In the previous chapter, we saw that protein flexibility is strictly connected to protein vibrations. In particular, low-frequency vibrations are the ones mostly affecting the flexibility of the structure. Also, we mentioned that protein flexibility has the potential to drive the structural rearrangements necessary for the conformational change. In this section, we will see how the low-frequency vibrations have been found to be strictly connected with the protein conformational change.

The first studies showing that the low-frequency protein motions are somehow connected with the observed biological protein mechanism were published in 1984 by Harrison [62] and in the following year by Brooks and Karplus [63]. In the first work, the author made use of a variational approach to calculate the protein normal modes and then applied the method to the case of yeast hexokinase, of which two conformations (open and closed form) were experimentally known. From a qualitative analysis of the obtained mode shapes, the author observed that some of the lowest-frequency modes of the open form (with wavenumber of about $7\ \mathrm{cm^{-1}}$) had strong components along the direction of the conformational change [62]. In the second work, the authors applied NMA to extract the fundamental mode shapes of lysozyme, and found that the lowest-frequency mode ($3.6\ \mathrm{cm^{-1}}$) represented well the hinge-bending motion which lysozyme was known to employ to perform its biological task [63].

More quantitative assessments of the connection between the protein conformational change and the low-frequency vibrations started to be carried out ten years later. In 1995, Marques and Sanejouand [64] performed all-atom NMA on the closed form of dimeric pig hearth citrate synthase and found that the lowest-frequency mode ($2.6\ \mathrm{cm^{-1}}$) compared well with the direction of the closed-to-open conformational change. In order to quantitatively assess this similarity, they made use of an overlap score O_n defined as:

$$O_n = \frac{|\boldsymbol{C}^{\mathrm{T}}\boldsymbol{\delta_n}|}{|\boldsymbol{C}^{\mathrm{T}}\boldsymbol{C}||\boldsymbol{\delta_n^{\mathrm{T}}\delta_n}|} = \frac{\left|\sum_{i=1}^{N}\left(c_{i,x}\delta_{n,i,x} + c_{i,y}\delta_{n,i,y} + c_{i,z}\delta_{n,i,z}\right)\right|}{\sqrt{\sum_{i=1}^{N}\left(c_i^2\right)}\sqrt{\sum_{i=1}^{N}\left(\delta_{n,i}^2\right)}}, \tag{4.6}$$

where $\boldsymbol{C}$ is the vector of conformational change displacements and $\boldsymbol{\delta_n}$ is the vector related to the nth mode shape. The overlap O_n defined in Equation (4.6) basically represents the normalized dot product between the displacement field of the conformational change $\boldsymbol{C}$ and the displacement field of the nth vibrational mode $\boldsymbol{\delta_n}$. When the nth mode shape is perfectly aligned to the conformational change, then O_n is equal to 1. When the two displacement fields are totally uncorrelated, i.e., orthogonal, O_n is equal to 0. Marques and Sanejouand [64] found that the overlap between the closed-to-open conformational change of citrate synthase and the first low-frequency mode calculated on the closed structure was 0.49. Although not exceptionally high, this value allowed to recognize that the lowest-frequency modes could somewhat be able to predict the directionality of the conformational change.

In the same year, Perahia and Mouawad [65] applied NMA with diagonalization in a mixed basis (DIMB) to extract the low-frequency modes of the T-state of hemoglobin. Then they compared the vibrational mode shapes to the conformational change of hemoglobin from the T-state to the R-state. The comparison was made by introducing another measure, the cumulative squared overlap CSO_n, which is defined as:

$$CSO_n = \sum_{k=1}^{n} O_k^2. \tag{4.7}$$

The CSO_n basically provides a numerical estimate of the contribution of the first n low-frequency modes to the conformational change. Due to the properties of the normal modes (which form a complete orthogonal basis), it is easy to see that CSO_n is equal to 1 when all the $3N$ modes are taken into account. Based on their calculations, Perahia and Mouawad [65] found that the two lowest-frequency modes alone were able to contribute to the T-R conformational change for the 70%, i.e., $CSO_2 \simeq 0.70$. From Figure 4.2 it can be seen that the second low-frequency mode (2.0 cm^{-1}) is the most involved into the T-R conformational change. There are also three other minor modes contributing to this transition, corresponding to frequencies of 1.7, 3.6, and 5.9 cm^{-1}. The displacements of the structure along these modes were found to mainly correspond to quaternary structural rearrangements. On the other hand, from Figure 4.2, it can be noticed that there exist numerous modes, with frequencies between 5.5 and 20 cm^{-1}, which only slightly contribute to the T-R transition individually, but whose aggregate contribution is not negligible. These motions were found to involve mainly tertiary structure rearrangements [65].

The first comprehensive investigation of the connection between the low-frequency modes and protein conformational changes was carried out by Tama and Sanejouand in 2001 [61]. In this work, the authors made use of coarse-grained ENMs to study the similarities between the conformational changes of twenty proteins and their low-frequency modes. From the results, it was observed that the direction of the conformational change is often well represented by one of the lowest-frequency vibrations. As an example, the second low-frequency mode of LAO-binding protein in its open form was found to closely match the observed open-to-closed transition, with an overlap O_2 of 0.84. Figure 4.3a shows the normalized values of the observed (thick line) and calculated (thin line) displacements. Overlap scores of 0.86 and 0.83 were also obtained when comparing the second mode of maltodextrin-binding protein and the third mode of citrate synthase to the corresponding open-to-closed conformational changes, respectively [61]. Other proteins also showed good overlap values for one of the low-frequency modes. Eventually, this was proof of the fact that the directions of the lowest-frequency modes often match the direction of the observed conformational change.

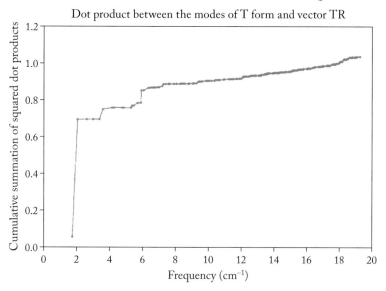

Figure 4.2: Conformational change of hemoglobin from T-state to R-state predicted by the low-frequency modes calculated on the T-state. The graph shows the *CSO*, defined in Equation (4.7), as a function of the mode frequency. Reprinted from Perahia, D. and Mouawad, L. Computation of low-frequency normal modes in macromolecules: Improvements to the method of diagonalization in a mixed basis and application to hemoglobin. *Comput. Chem.*, 19:241–246, 1995 [65]. Copyright 2021 with permission from Elsevier.

The work of Tama and Sanejouand [61] showed that the open-to-closed conformational change was easier to observe from some of the lowest-frequency modes of the open form than vice versa. For instance, the low-frequency modes of the closed forms of LAO-binding protein, maltodextrin-binding protein, and citrate synthase, led to maximum overlaps with the corresponding closed-to-open conformational changes of 0.40, 0.77, and 0.57, respectively—lower than the values obtained for the open-to-closed transitions. Another important result that Tama and Sanejouand [61] found from their analysis was that the low-frequency modes agree better with the observed conformational change when this transition is highly collective. That is, the higher the degree of collectivity κ of the conformational change, the higher our chances to extract low-frequency motions which are able to predict such transition. As we saw above, the open-to-closed transition of LAO-binding protein had a rather high degree of collectivity ($\kappa = 0.68$). As a result, for this protein we could observe a high similarity between the second low-frequency mode and the open-to-closed conformational change (Figure 4.3a, $O_2 = 0.84$). On the other hand, the conformational change of triglyceride lipase is highly localized, with $\kappa = 0.07$ (Figure 4.3b). For this protein, a maximum overlap of only 0.30 was found [61]. This suggested that the directions of localized transitions are not easy to describe in terms of the

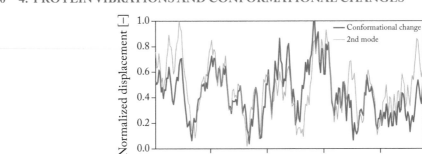

(a)

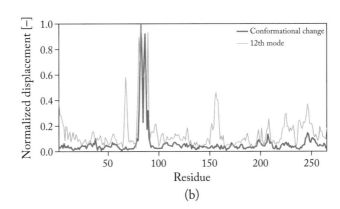

(b)

Figure 4.3: Comparison between the normalized displacements of conformational change (thick lines) and vibrational mode (thin lines) for: (a) LAO-binding protein—second low-frequency mode and (b) triglyceride lipase—twelfth low-frequency mode. Calculations run from some of the examples reported in Tama and Sanejouand [61].

lowest-frequency vibrations. Nevertheless, it was shown that the twelfth vibrational mode was still able to capture the amplitude distribution of the displacements involved into the transition, although not its directionality (Figure 4.3b). Therefore, protein vibrations can still carry some information regarding localized transitions as well.

In the following years, several studies have confirmed the previous findings, demonstrating that there exists a strict relationship between the low-frequency normal modes and protein conformational changes (Figure 4.4) [66]. Delarue and Sanejouand [67] investigated the open-to-closed transition in DNA polymerases and found that this can be well described with just a handful of the low-frequency modes extracted from the starting structure. In a comprehensive study, Krebs et al. [68] investigated 3,814 experimentally known protein motions and found that most of these can be described well by only a few low-frequency modes. In many cases, only one

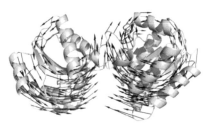

Figure 4.4: The observed conformational change (left) of Leu/Ile/Val-binding protein is accurately predicted by the second low-frequency mode (right), with an overlap $O_2 = 0.9$. Reprinted from Mahajan, S. and Sanejouand, Y. H. On the relationship between low-frequency normal modes and the large-scale conformational changes of proteins. *Arch. Biochem. Biophys.*, 567:59–65, 2015 [66]. Copyright 2021 with permission from Elsevier.

or two low-frequency vibrations are enough to describe accurately the observed protein motion [68]. Tobi and Bahar [69] analyzed the conformational changes of four proteins during protein-protein binding processes. Again, a good agreement was found between the vibrational modes calculated in the protein unbound form and the observed conformational change. These findings supported the idea that the native protein conformations are already predisposed to undergo the conformational fluctuations that are relevant to the biological function [69]. Similar results were found by Dobbins et al. [70] regarding protein-protein docking.

All these studies clearly showed that conformational changes, especially when they are highly collective, can often be described by using one or few among the low-frequency motions. However, we do not know *a priori* which mode is to be chosen. Is it the first one? The second? In 2006, Nicolay and Sanejouand [71] showed that the functional modes of the protein structure, i.e., the ones which are expected to carry relevant information for the biological function, are often among the most robust, where for robust modes they indicated those modes which are well conserved when the model details are changed [71]. Following the directions of these few robust modes, Mahajan and Sanejouand were also able to generate fairly accurate predictions of the observed conformational change [72]. A comprehensive study was then carried out by Yang et al. in 2008 [73], regarding the extent to which protein modes can be used to understand large-scale conformational changes. From the investigation of 170 pairs of open and closed protein conformations, it was found that protein conformational transitions usually fall into three categories: (i) the transitions whose direction can be explained well by low-frequency modes; (ii) the transitions that are not represented accurately by conventional ENM modes, but where adjustments based on the considerations and modeling of rigid clusters of atoms in the protein structure led to improved similarity; and (iii) the transitions whose intrinsic nature, e.g., the low degree of collectivity, prevents an accurate representation by means of the low-frequency vibrations [73]. From these considerations, it follows that low-frequency vibrations can indeed be helpful for the description of the more collective protein motions, especially when additional features of the protein structure are taken into account, such as the presence of rigid clusters. In the next section, we will describe a recent computational model (hdANM), which we developed together with the Jernigan's research group [60], in order to predict and visualize the large-scale collective motions of proteins, taking purposely into considerations the presence of rigid domains.

4.3   hdANM: A NEW TOOL FOR THE PREDICTION AND VISUALIZATION OF LARGE-SCALE HINGE MOTIONS

At the end of the previous chapter we briefly described the RTB [58], a method for the evaluation of protein vibrations based on the partition of the protein structure in rigid blocks. Based on this representation, the protein motion was simply described as the superimposition of the rigid translations and rotations of these blocks. This method was found to provide good predictions of the protein fluctuations and small-amplitude low-frequency motions [58]. We also mentioned that the method was further improved by Hoffmann and Grudinin [59], who proposed a methodology (NOLB) for the extrapolation of the RTB modes for the description of the large-scale nonlinear protein motions. The main advantage of NOLB over RTB is that the description of large-scale motion becomes more realistic, as the motions of blocks are described following their actual curvilinear pathway due to the rotations (Figure 4.5a) [59]. In this way, the large-scale extrapolation along the vibrational modes represent, in a better way, how a certain protein perform its function through its motion. Moreover, it does not give rise to those unrealistic stretches observed when the modes are extrapolated in a simple linear fashion (Figure 4.5b).

The main feature of RTB and NOLB is that they treat the entire protein as a union of translating and rotating rigid blocks. However, with this description, the highly flexible parts of the protein might not be well represented. On the other hand, most of the observed protein conformational transitions are known to involve the deformation of highly flexible portions accommodating the movements of quasi-rigid domains: this is especially true for the so-called hinge motions. In 1998, Gerstein and Krebs [74] carried out a comprehensive study, classifying most of the observed conformational changes into distinct categories. The two main classifications of the protein motion were shear motion and hinge motion (Figure 4.6).

The former is defined as a sliding mechanism of certain domains along the direction of the domain-domain interfaces (Figure 4.6, lower panel). The latter occurs when two (or more) domains are connected together by flexible hinges which allow the opening-closing motion of the domains (Figure 4.6, upper panel). It was found that most of the observed protein motions fall in the hinge-shear classification: 44% were classified as hinge motions and 14% as shear motions. The remaining 42% were found to fall into other categories, e.g., special mechanisms, motions involving partial refoldings, motions that cannot be classified, allosteric or non-allosteric motions, and complex motions [74]. Based on these findings, it is clear that hinge-domain motions

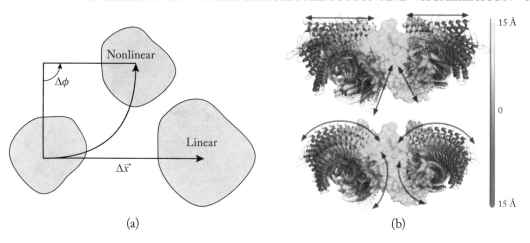

(a)   (b)

Figure 4.5: Large-scale nonlinear extrapolation of the vibrational modes in NOLB: (a) description of how following the rotation of the rigid block affects the final shape of the block and (b) linear (upper panel) and nonlinear (lower panel) large-scale motion of terminase pentase protein. Adapted with permission from Hoffmann, A. and Grudinin, S., NOLB: nonlinear rigid block normal mode analysis method. *J. Chem. Theory Comput.*, 13:2123–2134, 2017 [59]. Copyright 2021 American Chemical Society.

play an important role in most of the protein transitions. We recently developed a new computational model, the hinge-domain ANM (hdANM), for the prediction and visualization of these large-scale protein hinge motions [60].

The idea behind hdANM is fairly straightforward. As shown in Figure 4.6, in a protein structure you can often identify highly flexible portions, i.e., the hinges in a hinge motion or the interfaces in a shear motion, and two or more rigid portions, which represent the domains. These domains are supposed to move collectively as rigid bodies and the intra-domain fluctuations are assumed to be negligible. Hence, the hdANM starts from a simple coarse-grained ANM representation of the protein structure, i.e., a bunch of C^α atoms connected by Hookean springs (see Chapter 3), and then it also takes into account the conditions of rigid-body motion for the domain parts. Therefore, the protein is modeled as a mixture of both flexible parts (the hinges) and rigid blocks (the domains), as shown in Figure 4.7 [60]. The prediction of the hinge and domain regions in the protein structure is a subject widely addressed in the literature. A recent computational approach has been developed by Khade et al. [75], where the protein hinges are identified based on packing densities and alpha shapes, and validated by using permutation tests on B-factors.

Hinge motion

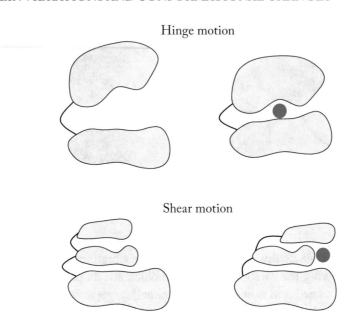

Shear motion

Figure 4.6: Schematic representation of shear and hinge protein motion.

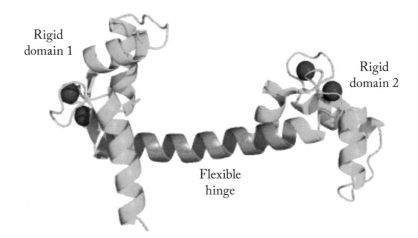

Figure 4.7: Hinge-domain partition for the open form of calmodulin (PDB: 1cll), as evaluated from the packing-based approach PACKMAN (https://packman.bb.iastate.edu) of Khade et al. [75]. Residues 67–88 form the central flexible hinge (in orange), whereas residues 4–66 and 89–147 form the two rigid domains (in green).

The fundamental equations of hdANM are obtained starting from the dynamic eigenvalue-eigenvector equations reported in the previous chapter, i.e.,

$$\left(H - \omega_n^2 M\right) \delta_n = 0, \tag{4.8}$$

where H is the $3N \times 3N$ ANM Hessian matrix, M the $3N \times 3N$ matrix containing the mass values of the amino acids, and ω_n^2 and δ_n the nth eigenvalue and eigenvector of the dynamical problem. Once the hinge-domain partition of the protein structure is known (Figure 4.7), the equations of rigid-body motion are then applied to the nodes belonging to the protein domains. In this way, the displacements of the domain nodes are only function of six DOFs, the three translations of the center of mass (COM) of the domain and the three rotations of the domain around its COM. In formulae:

$$\delta_x^i = \delta_x^d - \varphi_z^d \left(y_i - y_d\right) + \varphi_y^d \left(z_i - z_d\right), \tag{4.9a}$$

$$\delta_y^i = \delta_x^d + \varphi_z^d \left(x_i - x_d\right) - \varphi_x^d \left(z_i - z_d\right), \tag{4.9b}$$

$$\delta_z^i = \delta_x^d - \varphi_y^d \left(x_i - x_d\right) + \varphi_x^d \left(y_i - y_d\right), \tag{4.9c}$$

where δ_x^i, δ_y^i, and δ_z^i are the X-, Y-, and Z-components of the displacement of the ith node (with coordinates x_i, y_i, and z_i) belonging to domain d, δ_x^d, δ_y^d, and δ_z^d are the X-, Y-, and Z-components of the displacement of the COM of domain d (with coordinates x_d, y_d, and z_d), and φ_x^d, φ_y^d, and φ_z^d are the X-, Y-, and Z-components of the rotation of domain d around its COM. Equations (4.9) establish the fact that each domain d translates and rotates as a rigid body without any internal deformation. Coupling Equations (4.9) to the dynamical problem formulated in Equation (4.8), one obtains the fundamental hdANM eigenvalue-eigenvector equation [60]:

$$\left(H' - \omega_n'^2 M'\right) \delta_n' = 0, \tag{4.10}$$

where H' and M' are the new Hessian (stiffness) and mass matrices in the hdANM framework, and $\omega_n'^2$ and δ_n' the corresponding eigenvalues and eigenvectors. Notice that, due to the rigidity of the domains expressed in Equation (4.9), the total number of DOFs in the hdANM (and therefore the total dimension of the H' and M' matrices) is $6D + 3n_h$, D being the total number of domains and n_h the total number of nodes contained in the flexible regions. Notice that $6D + 3n_h$ is usually much lower than the number of DOFs of the starting ANM, i.e., $3N$. As an example, the open structure of calmodulin reported in Figure 4.7 counts 144 residues, therefore the C^α-based coarse-grained ANM would involve $3 \times 144 = 432$ DOFs. Conversely, considering the structure made up of the two green rigid domains reported in the figure, i.e., $D = 2$, and a total number of residues in the flexible hinge region $n_h = 22$ (residues 67–88), one obtains a total number of DOFs in the hdANM equal to $6 \times 2 + 3 \times 22 = 78$—much lower than the 432 DOFs taken into account in the ANM. This leads to a faster diagonalization of the Hessian and mass matrices, and therefore to an enhanced computational speed.

Once the $(6D + 3n_h) \times 1$ hdANM eigenvectors δ'_n are obtained from Equation (4.10), these can be used to extrapolate the protein motions. This can be done in two ways. First, the eigenvectors δ'_n can be directly transformed into displacement vectors with dimension $3N \times 1$, by applying the Equations (4.9) to the DOFs associated with the domains. Then, these $3N \times 1$ vectors can be directly used to visualize the motion via a simple linear scaling, as usually done in ANM and RTB. In the second way, we can apply a methodology similar to the NOLB and follow the rotations of the rigid domains in order to find the curvilinear, i.e., nonlinear, pathways (see Figure 4.5). This can be done by writing the displacement vector of each ith node belonging to domain d, δ^i, as a linear function of the vector associated with the translations of the COM, δ^d, and as a nonlinear function of the vector associated with the rotational components of the domain, φ^d:

$$\delta^i (S) = S\delta^d + R \left(S\varphi^d \right) r_i, \qquad (4.11)$$

where S is the scale factor associated with each intermediate conformation, r_i is the vector of the coordinates of node i, and R is the rotation matrix which takes into account the rigid-body rotation of the domain based on the scaled value $S\varphi^d$ of the rotation [60]. By applying Equation (4.11) to the nodes of the rigid domains, we can visualize the large-scale curvilinear motion of the domains, similar to the NOLB approach, without generating those unrealistic internal stretches due to the large-scale linear extrapolations (see Figure 4.5).

Via the hdANM, we were able to investigate a variety of conformational transitions elucidating the opening and closing mechanisms observed in various proteins [60]. As an example, Figure 4.8 shows the predicted motions with the nonlinear extrapolation along the three lowest-frequency modes for calmodulin. As can be seen, these motions involve high deformations in the hinge parts (orange portions), whereas the domains (green portions) experience pronounced rigid-body motions. The combination of the specific hinge deformation and the directions of domain motions can lead to the identification of a variety of protein mechanisms, such as hinge twisting and hinge bending in different directions (Figure 4.8). The knowledge of these types of motion can ultimately provide a deeper comprehension of the specific protein mechanisms required for a correct functionality [60].

The hdANM can also be seen as the ultimate generalization of the previous elastic models. In fact, when we define no domains in the protein ($D = 0$), therefore the entire protein structure is treated as a flexible hinge ($n_h = N$), the hdANM basically overlaps with the conventional ANM, and the total DOFs of the system are simply $3N$. On the other hand, when we have no flexible regions ($n_h = 0$), but the protein is modeled just as a union of D rigid domains, the hdANM overlaps with the RTB and the DOFs are now $6D$. Conversely, when you consider both flexible and rigid portions in your protein structure (see Figure 4.7), you come up with a new elastic model which is something in between and more comprehensive than the traditional ANM and RTB approaches.

In this and the previous chapter, we learned how simplified modeling techniques, relying on Normal Mode Analysis and elastic models, can help you to get insights into protein dynam-

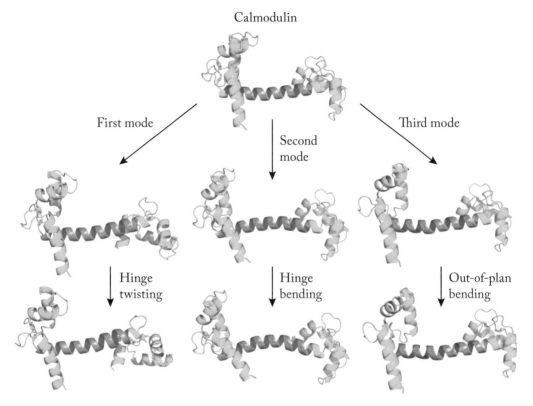

Figure 4.8: Prediction and visualization of hinge protein motions from hdANM (https://hdanm.bb.iastate.edu). Example for calmodulin (1cll) for the first, second, and third low-frequency mode. Each motion provides a specific mechanism based on the hinge deformation and domain rigid motions. Domains are in green while the hinge is in orange, as in Figure 4.7.

ics and flexibility, and their connection with the observed conformational transitions. In the following chapter, we will see that some insights regarding protein dynamics and its relationship with the biological function can also be retrieved by means of experimental techniques. In particular, by means of spectroscopy techniques such as Raman and Terahertz Time Domain Spectroscopy (THz-TDS). Again, we will see how the low-frequency vibrations are found to involve collective motions, often carrying biologically relevant information.

CHAPTER 5

Exploring Protein Vibrations Experimentally: Raman and THz-TDS

In the previous chapters we focused our attention on the computational approaches to unravel protein vibrations, flexibility, and their connection with the biological functionality. In this last chapter, we will describe experimental techniques used to detect and investigate protein vibrations based on spectroscopic approaches. In general terms, spectroscopy deals with the analysis of the interaction between light and matter, as a function of the frequency of the incident light. Different spectroscopy techniques differ depending on the energy of the incident radiation (visible, infrared, terahertz, etc.) and on the phenomena underlying the light-matter interaction (absorption, emission, scattering, etc.). In this chapter, we will focus on two spectroscopy methods for protein analysis, namely Raman spectroscopy and terahertz time-domain spectroscopy (THz-TDS).

5.1 RAMAN SPECTROSCOPY AND APPLICATION TO PROTEINS

Raman spectroscopy was developed based on the discovery of the Raman effect by the Indian physicist Chandrasekhara Venkata Raman in the 20th century [76]. In this context, a certain material is irradiated by a monochromatic light beam (usually a laser in the visible, near-infrared or near-ultraviolet range), and the scattered light is collected by a detector after the interaction with the sample.

It was observed that the scattered light shows different frequency components (Figure 5.1). The larger component usually corresponds to the same frequency of the incident light, and this is referred to as the Rayleigh scattering. Other components can also be detected at lower or higher frequencies. These are due to the Raman scattering effect and are associated with the Stokes and anti-Stokes shifts, respectively (Figure 5.1). These shifts are informative of the vibrational states of the sample under investigation and can be used to identify the material and its internal properties.

There is a relatively simple explanation of the Raman effect, based on the classical theory of vibrational modes and their interaction with an incident electromagnetic field. Imagine

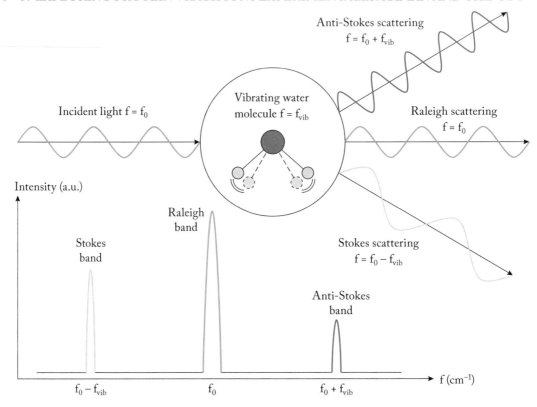

Figure 5.1: Raman spectroscopy: Rayleigh, Stokes, and anti-Stokes components resulting from the interaction between the incident light and the vibrational activity of the material.

to irradiate a molecule with an oscillating electric field E (Figure 5.1), whose time-dependent variation is described by:

$$E = E_0 \cos (2\pi f_0 t),\qquad(5.1)$$

E_0 and f_0 being the amplitude and frequency of the incident field, respectively. As soon as the electric field interacts with the molecule, this one becomes polarized and a dipole moment μ arises:

$$\mu = \alpha E = \alpha E_0 \cos (2\pi f_0 t),\qquad(5.2)$$

where α is the polarizability of the molecule, which is a measure of the extent to which the electric charges within the molecule can be displaced by an external electric field. In general terms, the polarizability α depends on the vibrational state of the molecule, i.e.,

$$\alpha = \alpha_0 + \alpha' \cos (2\pi f_{vib} t),\qquad(5.3)$$

where α' represents the rate of change of polarizability with the vibration and f_{vib} the frequency associated with the vibrational state of the molecule. Putting together Equations (5.2) and (5.3), one obtains the total induced dipole moment μ:

$$\mu = \alpha_0 E_0 \cos(2\pi f_0 t) + \alpha' E_0 \cos(2\pi f_0 t) \cos(2\pi f_{vib} t). \tag{5.4}$$

Exploiting trigonometric relationships, the second term of the r.h.s. of Equation (5.4) can be divided into two terms, leading to:

$$\mu = \alpha_0 E_0 \sin(2\pi f_0 t) + \frac{\alpha' E_0}{2} \cos[2\pi (f_0 - f_{vib}) t]$$
$$+ \frac{\alpha' E_0}{2} \cos[2\pi (f_0 + f_{vib}) t]. \tag{5.5}$$

Due to the fact that the intensity of the collected light is proportional to the total dipole moment of the molecules, the three terms reported in Equation (5.5) represent the three collected signals represented in Figure 5.1. The first term stands for the Rayleigh scattering (the frequency of this component is equal to the frequency of the incident light f_0). The second term represents the Stokes shift (the frequency of the collected light is lower than the frequency of the incident light f_0 and the shift matches the vibrational frequency of the molecule f_{vib}). The third term represents the anti-Stokes shift (the frequency of the collected light is higher than the frequency of the incident light f_0 and the shift again matches the vibrational frequency of the molecule f_{vib}). From a direct analysis of Equation (5.5), it is quite clear that Stokes and anti-Stokes shift are only obtained if α' is different from zero for the considered vibrational state. From this consideration, the selection rule for Raman activity directly follows. Only those vibrations involving a change in the overall molecule polarizability α' are Raman-active and therefore are able to cause Raman scattering.

As we have seen, Stokes and anti-Stokes bands in the Raman plots are informative of the vibrational states of a certain sample under investigation (Figure 5.1). Commonly, Raman spectra are centered at the frequency of the incident light f_0, and therefore the positions of the Stokes and anti-Stokes bands are directly identified by means of the positive and negative shifts from the laser frequency. Moreover, since Stokes bands usually exhibit higher intensities, these are generally assigned positive frequency shifts and are the ones more used for material identification and analysis. The position, intensity value, and linewidth of these Raman bands are informative of different aspects of the material under scrutiny. The position of the band is a direct information of the frequency of vibration of the irradiated molecule or molecular groups, f_{vib}. This often allows to identify specific chemical groups as well as their structural arrangement within a sample. The intensity of the Raman band is often associated with the concentration of the corresponding chemical group within the sample. Finally, the linewidth of the band is often a fingerprint of the structural disorder. Broader bands are usually associated with more disordered configurations, whereas narrower peaks to more ordered configurations.

Raman spectroscopy has been deeply used in the last decades for material identification and analysis in a variety of fields, ranging from biotechnology, food and drug control, pharmaceutics, material sciences, etc. It has many advantages among other spectroscopy techniques, such as non-destructiveness, very little or no need for sample preparation, possibility to probe materials *in situ*, etc. Among other materials, proteins have also been extensively probed by Raman spectroscopy in order to extract information about their structure, function and specific features [77]–[80].

Typical applications of Raman to protein analysis include the identification and interpretation of bands associated with secondary structures, the assessment of the protein conformational state and the assignment of specific amino acid vibrational bands [79]. The most characteristic bands in the Raman spectra of proteins are those associated with the vibrations of the CONH group in the protein backbone. These are known as the amide bands, and are differentiated depending on the specific vibration involved [78]. Amide A is found to occur at about 3500 cm^{-1} (105 THz) and involves the NH stretching. Amide B involves a similar vibration and occurs at about 3100 cm^{-1} (93 THz). Among the other amide bands (amide I to VII), the first three are of main importance. Amide I occurs at 1600–1690 cm^{-1} (48–51 THz) and involves the stretching of the CO bond, while amide II and III occur at 1480–1580 cm^{-1} (44–47 THz) and 1230–1300 cm^{-1} (37–39 THz), respectively, and both are associated with coupled CN stretching and NH bending vibrations of the peptide group [78]. Other important features that can be observed and analyzed in protein Raman spectra are associated with amino acid side-chain vibrations, e.g., the tryptophan and tyrosine doublets at $1340/1360$ cm^{-1} (40.5 THz) and $833/860$ cm^{-1} (25.5 THz), respectively. Sulphur-containing residues can also be clearly identified in Raman spectra [78]. As an example, Figure 5.2 reports the Raman spectrum of a lyophilized powder of Na/K-ATPase, showing various Raman peaks whose assignment is reported in Table 5.1 [81]. As can be seen, all these Raman bands are informative of a particular vibrational state of the protein structure.

Despite some intrinsic challenges related to the application of Raman spectroscopy to biological materials such as proteins, plenty of information is now available on the Raman spectra of proteins [77]–[80]. However, little attention has been paid to the low-frequency part of the spectrum, i.e., in the sub-500 cm^{-1} range, which we know from the previous chapters to be the most important part of the frequency spectrum as far as the overall protein flexibility and its connection to the biological mechanism is concerned. This is mainly due to the fact that the low-frequency part of the spectrum is highly hindered by the Rayleigh scattering component. Nevertheless, some studies have been carried out in order to explore this part of the spectrum thanks to some specific technological adjustments.

One of the first detections of low-frequency Raman bands in proteins was achieved by Brown et al. [82] in 1972, by using a iodine filter technique and a double-grating monochromator. They found a pronounced Raman peak at 29 cm^{-1} (0.87 THz) for α-chymotrypsin samples prepared in several ways. Interestingly, this low-frequency Raman band was found for all protein

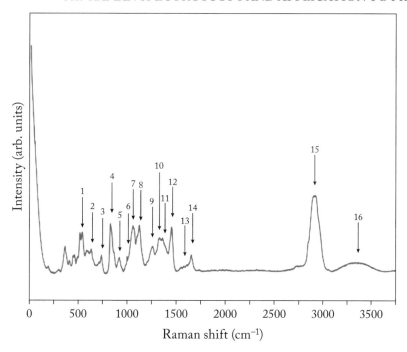

Figure 5.2: Example of Raman spectrum with identified peaks in the 500–3500 cm^{-1} range for lyophilized Na/K-ATPase. The chemical assignment of the peaks is reported in Table 5.1. Reprinted from Lacidogna et al. Raman spectroscopy of Na/K-ATPase with special focus on low-frequency vibrations. *Vibr. Spectr.*, 92:298–301, 2017 [81]. Copyright 2021 with permission from Elsevier.

samples, except for that denatured with sodium-dodecyl-sulfate (SDS). This suggested that such low-frequency vibration must arise from vibrations that involve all, or very large portions, of the three-dimensional protein structure [82]. Few years later, Genzel et al. [83] performed Raman spectroscopy on lysozyme samples, both in crystal forms and aqueous solutions. They found Raman peaks for the crystal samples at 25, 75, 115, and 160 cm^{-1}, whereas only the bands at 75 and 160 cm^{-1} were found again in the aqueous solution. From these results, it was argued that the low-frequency peak at 25 cm^{-1} might arise from inter-molecular vibrations between the lysozyme molecules when arranged in the crystal lattice [83]. A comprehensive study was carried out by Painter et al. [84], where various low-frequency Raman bands were observed for different proteins (see Table 5.2).

More recently, Kalanoor et al. [85] developed a method based on volume holographic filters with a single-stage spectrometer, to extract the low-frequency Raman modes of biomolecules. From the application of the method, Raman bands below 50 cm^{-1} could be iden-

Table 5.1: Assignment of peaks reported in the Raman spectrum of Na/K-ATPase (Figure 5.2). Reprinted from Lacidogna et al. Raman spectroscopy of Na/K-ATPase with special focus on low-frequency vibrations. *Vibr. Spectr.*, 92:298–301, 2017 [81]. Copyright 2021 with permission from Elsevier.

N	Raman Shift (cm^{-1})	Assignment	N	Raman Shift (cm^{-1})	Assignment
1	524	S–S stretching	9	1260	Amide III
2	640	C–C twisting in Tyr	10	1335	CH$_3$–CH$_2$ wagging
3	740	C–S stretching	11	1365	Trp
4	831	Tyr	12	1454	C–H bending
5	930	C–C stretching	13	1585	Tyr, Phe
6	1004	Phe	14	1620–1670	Amide I
7	1067	Pro	15	2900–3000	CH$_3$–CH$_2$ stretching
8	1125	Trp	16	3200–3600	H$_2$O stretching

Table 5.2: Low-frequency Raman peaks in proteins [84]

Protein	Observed Raman Line (cm^{-1})	Protein	Observed Raman Line (cm^{-1})
Insulin	22	Concanavalin A	20
Lysozyme	25	BSA*	14
β-lactoglobulin	25	BIgG**	28
α-chymotrypsin	29	Adolase	32
Pepsin	20	Thyroglobulin	17
Ovalbumin	22		

*Bovine serum albumin; **Bovine immunoglobulin G.

tified for several proteins, such as lysozyme, bovine serum albumin (BSA) and immunoglobulin G (IgG). Recently, our research group also performed Raman measurements on protein samples with special focus on the low-frequency range [51, 81]. Samples of lysozyme and Na/K-ATPase were analyzed via Raman lasers equipped with special ultra-low-frequency (ULF) filters, which enabled to detect Raman bands down to 5 cm^{-1}. Figure 5.3 shows the obtained Raman spectra for lysozyme and Na/K-ATPase, which highlight clear peaks in the low-frequency range. A strong peak at 28 cm^{-1} (0.84 THz) was found for lysozyme (Figure 5.3a), whereas a peak at 27 cm^{-1} (0.81 THz) for Na/K-ATPase (Figure 5.3b).

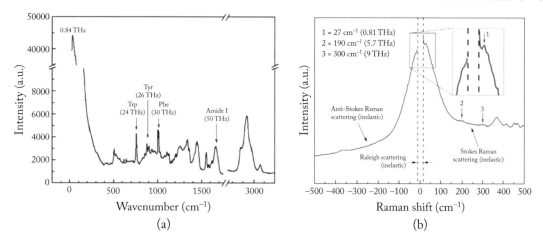

Figure 5.3: Raman spectrum of: (a) crystallized hen egg-white lysozyme and (b) lyophilized powder of Na/K-ATPase. Spectra obtained by means of ULF filter in order to focus on the low-frequency region of the Raman spectrum. (a) Reprinted from Carpinteri A. et al. Terahertz mechanical vibrations in lysozyme: Raman spectroscopy vs. modal analysis. *J. Mol. Struct.*, 1139:222–230, 2017, with permission from Elsevier [51]. (b) Reprinted by permission of Taylor & Francis from Carpinteri A. et al. Terahertz vibration modes in Na/K-ATPase. *J. Biomol. Struct. Dyn.*, 37:256–264, 2019 [52].

It remains to unarguably clarify the nature of these low-frequency Raman peaks. Where do they arise from? What do they really represent? It is not so easy to answer to these questions with the knowledge of today. We have seen in Chapter 3, that higher frequencies are usually associated with small-scale vibrations, whereas lower frequencies often involve large-scale collective movements. In the realm of proteins, this means that low-frequency vibrations should usually be associated with global collective motions of the overall protein structure, whereas high-frequency vibrations should represent the localized motions of small protein portions or single chemical groups. In a way, the Raman peak assignments reported in Table 5.1 show that higher frequencies usually involve localized vibrations of smaller chemical groups, although this is not always the case.

When it comes to the lowest-frequency Raman peaks, things are more difficult to explain because of different reasons. First, in Chapter 3 we have already seen that the specific frequencies extracted from the various computational approaches (all-atom NMA, coarse-grained NMA, ENMs, FE models, etc.) are strongly dependent on the model details and should be treated carefully. Second, we have mentioned above that the selection rule for Raman vibrational activity says that only those vibrational modes which generate a change in the molecule polarizability α' are Raman-active. Therefore, an association between a certain low-frequency Raman peak and a certain low-frequency vibrational mode should also investigate the change in polarizability of such a mode. Finally, in Chapter 3 we saw that the density of the protein vibrational modes is very populated in the low-frequency part of the spectrum (see Figure 3.1). This potentially prevents a reliable one-on-one assignment between a certain low-frequency Raman band and a specific vibrational mode.

In the previous chapters we saw that the lowest-frequency vibrations of proteins obtained by computational techniques were found to occur down to 1 cm^{-1} (30 GHz). Future research efforts in enhancing the resolution of low-frequency Raman measurements, coupled with multiphysics normal mode analysis (also including calculations of the protein polarizability), might lead us to a better understanding of the low-frequency protein vibrations, both from a computational and experimental perspective.

5.2　THz-TDS AND APPLICATION TO PROTEINS

Another technique that has found more and more application in the detection of protein dynamics since the last two decades is terahertz time-domain spectroscopy (THz-TDS). THz-TDS is a technique that relies on the generation and detection of photons with a frequency in the THz range. For a long time, there was a gap in the electromagnetic spectrum, approximately in the region between 0.1 and 1.5 THz (3–50 cm^{-1}), which was difficult to detect due to the lack of proper technological devices. This region lies in the mid- and far-infrared part of the spectrum (MIR–FIR) and it has been known for a long time as the "terahertz gap." With the advance of new emitters and detectors, such as photoconductive antennae (PCA), this region of the spectrum became finally accessible.

Figure 5.4 shows a sketch representation of a THz-TDS device operating in transmission mode. A femtosecond laser generates a pulse which is split into two separate beams by a beam splitter (BS). One beam is used to generate the THz radiation modulated with a certain frequency f (lower beam in Figure 5.4). After various reflections this beam hits the investigated sample within a purge-box which is usually purged with dry air or nitrogen. The other beam is sent to a delay line (upper beam in Figure 5.4), and finally used to detect the emitted THz signal in the time-domain $E(t)$ (Figure 5.5a) [86].

The measured THz signal can then be transformed in the frequency-domain spectrum $E(\omega)$ via Fourier Transform (FT):

$$E(\omega) = \frac{1}{\sqrt{2\pi}} \int_{-\infty}^{+\infty} E(t)\, \mathrm{e}^{-\mathrm{i}\omega t}\, dt. \tag{5.6}$$

The complex spectrum $E(\omega)$ carries the information about the intensity $I(\omega)$ and phase $\varphi(\omega)$, i.e.,

$$E(\omega) = I(\omega)\, \mathrm{e}^{\mathrm{i}\varphi(\omega)}, \tag{5.7}$$

which can finally be used to directly calculate the complex-value refraction index $n(\omega)$ of the material as a function of ω [86].

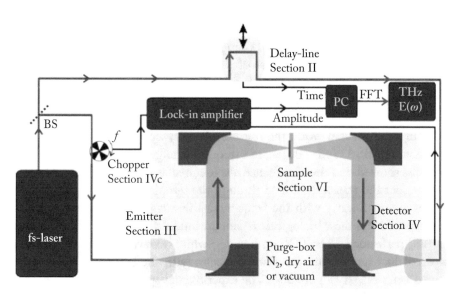

Figure 5.4: Sketch representation of THz-TDS device. Reprinted from Neu, J. and Schmuttenmaer, C. A. Tutorial: An introduction to terahertz time domain spectroscopy (THz-TDS). *J. Appl. Phys.*, 124:231101, 2018 [86], with the permission of AIP Publishing.

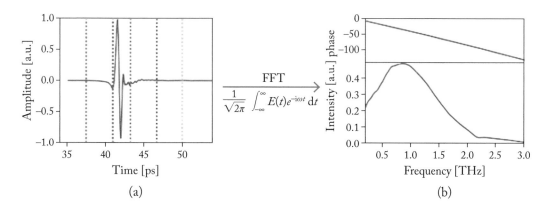

Figure 5.5: THz signal: (a) signal in the time-domain and (b) signal in the frequency-domain (separation of intensity and phase). Reprinted from Neu, J. and Schmuttenmaer, C. A. Tutorial: An introduction to terahertz time domain spectroscopy (THz-TDS). *J. Appl. Phys.*, 124:231101, 2018 [86], with the permission of AIP Publishing.

Throughout this book, we have seen that protein dynamics is pivotal to its biological functioning. Moreover, it emerged that low-frequency vibrations are key players in this process, as they are able to drive conformational changes and are the most responsible for the protein flexibility. We have also seen that most of these vibrations are found to occur in the THz and sub-THz frequency range. THz-TDS seems then the perfect experimental technique to see what happens to proteins in this frequency range [87].

The first study applying THz-TDS to protein samples was carried out by Markelz et al. [88] in 2000. In that work, the researchers analyzed lyophilized powders of DNA, bovine serum albumin (BSA) and collagen, in a frequency range between 0.06 and 2 THz (2–67 cm^{-1}). The results showed that the absorbance abs (defined as $abs = -\log(E_{t,s}/E_{t,r})$, where $E_{t,s}$ and $E_{t,r}$ represent the transmitted field through the sample and through a reference, respectively) increases almost linearly with the frequency. This suggested that a large number of the low-frequency modes of these biological systems are indeed IR active. In a subsequent work, Markelz et al. [89] performed THz-TDS on hen egg-white lysozyme and horse hearth myoglobin and compared the experimental absorbance spectra with the computed spectra of normal modes obtained from NMA (Figure 5.6). The experimental spectra (Figures 5.6a and 5.6b) reflected the density of the low-frequency vibrational modes (Figure 5.6c and 5.6d), suggesting that THz-TDS can be useful to probe the overall protein dynamics and flexibility [89].

THz-TDS measurements were also performed on thick films of bacteriorhodopsin, a photoreceptor protein which is known to undergo a conformational change due to a retinal 13-cis isomerization upon illumination with a light beam at 570 nm [89]. Figure 5.7a shows the THz absorbance spectrum of the protein film under different conditions. The continuous line refers to the spectrum of the protein in the ground state at 295 K. The dotted line indicates the ground state of the protein at 233 K. It is evident how cooling the sample reduces the flexibility of the protein, i.e., the density of the low-frequency normal modes, and therefore the absorbance spectrum gets reduced. The dashed line refers to the protein at 233 K in the M state, i.e., when the conformational change has been triggered upon illumination at 570 nm. As can be seen, the absorbance spectrum changes its shape, indicating that a protein conformational change can indeed be observed through THz-TDS measurements [89]. Finally, the dot-dashed line refers to the ground state of the protein at 295 K after warming and switching the illumination off. These results suggested for the first time that THz-TDS might be a useful technique to monitor protein conformational changes through the analysis of THz absorbance spectra. This was also found in a subsequent work by Chen et al. [90], where the absorbance spectrum of hen egg-white lysozyme was found to get modified when the protein is bound to the inhibitor triacetyl-glucosamine (3NAG).

In [89] the authors also compared the absorbance spectra of wild-type bacteriorhodopsin and of the D96N mutant, where the aspartic acid in position 96 was replaced by an asparagine residue. Again, a change in the spectrum was observed (Figure 5.7b), suggesting that the mutant is less flexible than the wild-type [89]. THz absorbance spectra were then proven also to be able

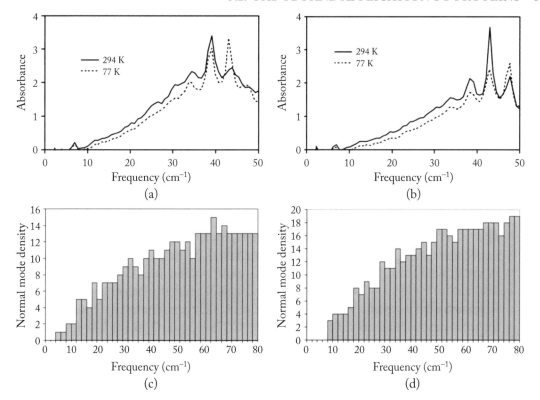

Figure 5.6: THz-TDS absorbance spectra for: (a) hen egg-white lysozyme and (b) horse hearth myoglobin. Computed normal mode density for: (c) hen egg-white lysozyme and (d) horse hearth myoglobin. Reprinted with permission from Markelz, A. et al. THz time domain spectroscopy of biomolecular conformational modes. *Phys. Med. Biol.*, 47:3797–3805, 2002 [89]. © Institute of Physics and Engineering in Medicine. Reproduced by permission of IOP Publishing. All rights reserved.

to provide useful insights about the effect of mutations. This was further confirmed by a more comprehensive research work by Whitmire et al. [91] in the following year.

In the above examples, we saw how protein flexibility can be assessed in terms of THz absorbance spectra. However, in Chapter 4 we have also seen how specific low-frequency modes are very informative of the protein functionality. Therefore, the question arises whether it is possible to identify single modes by THz-TDS. Significant attempts were carried out by Balu et al. [92], where the experimental THz absorbance spectra of rhodopsin and bacteriorhodopsin were compared to theoretical spectra, instead of the density of normal modes (Figure 5.6). The

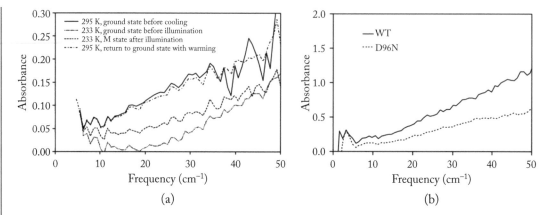

Figure 5.7: THz-TDS absorbance spectra for: (a) bacteriorhodopsin in the ground state at 295 K (continuous line), ground state at 233 K (dotted line), M state at 233 K (dashed line), and again in the ground state at 295 K after warming and switching the illumination off (dot-dahsed line) and (b) wild-type (continuous line) and D96N mutant (dashed line) bacteriorhodopsin. Reprinted with permission from Markelz, A. et al. THz time domain spectroscopy of biomolecular conformational modes. *Phys. Med. Biol.*, 47:3797–3805, 2002 [89]. © Institute of Physics and Engineering in Medicine. Reproduced by permission of IOP Publishing. All rights reserved.

theoretical intensity of the THz spectrum related to each mode i was calculated as:

$$I_i \cong \sum_{j=1}^{3N} \left| \frac{e_j^0 \delta_{i,j}}{\Delta Q_i} \right|^2, \tag{5.8}$$

where the sum is performed over all the $3N$ mass-weighted modal displacements $\delta_{i,j}$, e_j^0 is the static charge value associated with each atom coordinate and ΔQ_i represents the total root-mean-square atomic displacement for mode i [92]. Since Equation (5.8) provides an integrated intensity, a Lorentzian function to describe each line shape was used. Refer to [92] for more details. The analysis of the vibrational modes providing the highest computed intensity provided some information about the localization of the motion in these two proteins within different regions of the THz spectrum [92]. However, the approach did still not provide a reliable method for a clear one-by-one mode identification.

A few years later, Acbas, G. et al. [93] developed a new methodology based on THz-TDS to detect single-mode protein vibrations. The method was named crystal anisotropy terahertz microscopy (CATM), as it was based on measuring the difference in the THz absorbance spectrum of a protein crystal between a reference position and a variable orientation ϑ. In this way, the isotropic component of the absorbance due to the librational motions of the solvent and peptide chains could be removed, and only the information about the dipole variation related

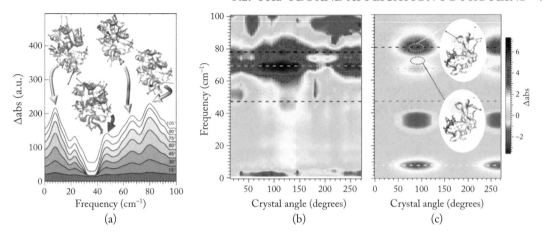

Figure 5.8: (a) Calculated Δ*abs* for different crystal orientations of chicken egg-white lysozyme. Individual motions from NMA calculations are reported close to the Δ*abs* peaks. Reprinted by permission from Springer from Acbas, G. et al. Optical measurements of long-range protein vibrations. *Nat. Commun.*, 5:1–7, 2014 [93]. (b) CATM Δ*abs* spectrum for hydrated chicken egg-white lysozyme, compared with (c) the calculated spectrum, where vector motions at 72 and 80 cm^{-1} (2.16 and 2.4 THz) are shown. Reprinted by permission from Springer from Niessen, K. A. et al. Terahertz optical measurements of correlated motions with possible allosteric function. *Biophys. Rev.*, 7:201–216, 2015 [94].

to each vibrational motion and specific light orientation was maintained. By taking the angular difference in the THz absorbance spectra, Δ*abs*, the isotropic background can then be removed and the identification of individual structural motions becomes possible. See [93] for more details.

Figure 5.8a shows the calculated Δ*abs* spectrum for chicken egg-white lysozyme up to 100 cm^{-1} (3 THz). Peaks in the Δ*abs* spectrum at 8, 47, 69, and 80 cm^{-1} are representative of individual protein vibrations and the corresponding displacement vectors extracted from NMA are reported close to each corresponding peak.

Figure 5.8b also compares the experimental and computed CATM spectra, where the value of Δ*abs* is shown in color scale for each frequency value and crystal orientation [94]. As can be seen, similarities arise between the two spectra, suggesting that CATM can indeed be useful to perform single-mode identification in THz spectra [93, 94]. More recently, CATM methodology has also been applied to investigate the variation of motion directionality upon ligand-binding [95]. The technique was carried out to analyze sample of chicken egg-white lysozyme both in the free state and bound with the inhibitor 3NAG. The change in the CATM spectrum was much more evident than the difference in the classical THz absorbance spectrum [90], thus suggesting that CATM can be a promising technology for the individuation of protein motion directionality as well [95].

All these recent studies suggest that CATM, and THz-TDS in general, coupled with computational techniques such as NMA, have a huge potential in deciphering protein vibrations in the THz frequency range. Hopefully, future research efforts will lead us to further knowledge about this important subject.

Bibliography

[1] Carpinteri, A. *Advanced Structural Mechanics*, CRC Press, Taylor & Francis Group, Boca Raton, 2017. DOI: 10.1201/9781315375298. 4, 39

[2] Yoshida, S. *Waves: Fundamentals and Dynamics*, Morgan & Claypool, San Rafael, CA, 2017. DOI: 10.1088/978-1-6817-4573-2. 5

[3] Alberts, B., Johnson, A., Lewis, J., Morgan, D., Raff, M., Roberts, K., and Walter, P. *Molec. Biol. Cell*, 6th ed., Garland Science, New York, 2002. 8, 14

[4] Bahar, I., Jernigan, R. L., and Dill, K. A. *Protein Actions: Principles and Modeling*, Garland Science, New York, 2017. 8, 14

[5] Eisenberg, D. The discovery of the α-helix and β-sheet, the principal structural features of proteins, *Proc. Natl. Acad. Sci.*, 100:11207–11210, 2003. DOI: 10.1073/pnas.2034522100. 11

[6] Pauling, L., Corey, R. B., and Branson, H. R. The structure of proteins: Two hydrogen-bonded helical configurations of the polypeptide chain, *Proc. Natl. Acad. Sci.*, 37:205–211, 1951. DOI: 10.1073/pnas.37.4.205. 12

[7] Pauling, L. and Corey, R. B. The pleated sheet, a new layer configuration of polypeptide chains, *Proc. Natl. Acad. Sci.*, 37:251–256, 1951. DOI: 10.1073/pnas.37.5.251. 12

[8] Huang, D. M. and Chandler, D. Temperature and length scale dependence of hydrophobic effects and their possible implications for protein folding, *Proc. Natl. Acad. Sci.*, 97:8324–8327, 2000. DOI: 10.1073/pnas.120176397. 16

[9] Anfinsen, C. B. The formation and stabilization of protein structure, *Biochem. J.*, 128:737–749, 1972. DOI: 10.1042/bj1280737. 16

[10] Anfinsen, C. B. Principles that govern the folding of protein chains, *Science*, 181:223–230, 1973. DOI: 10.1126/science.181.4096.223. 16

[11] Levinthal, C. Are there pathways for protein folding?, *J. Chim. Phys.*, 1968. DOI: 10.1051/jcp/1968650044. 17

[12] Levinthal, C. How to fold graciously, *Mossbauer Spectroscopy in Biological Systems*, University of Illinois Press, 1969. 17

[13] Leopold, P. E., Montal, M., and Onuchic, J. N. Protein folding funnels: A kinetic approach to the sequence-structure relationship, *Proc. Natl. Acad. Sci.*, 89:8721–8725, 1992. DOI: 10.1073/pnas.89.18.8721. 17

[14] Onuchic, J. N., Luthey-Schulten, Z., and Wolynes, P. G. Theory of protein folding: The energy landscape perspective, *Annu. Rev. Phys. Chem.*, 48:545–600, 1997. DOI: 10.1146/annurev.physchem.48.1.545. 17

[15] Rounsevell, R., Forman, J. R., and Clarke, J. Atomic force microscopy: Mechanical unfolding of proteins, *Methods*, 34:100–111, 2004. DOI: 10.1016/j.ymeth.2004.03.007. 18

[16] Go, N. Theoretical studies of protein folding, *Annu. Rev. Biophys. Bioeng.*, 12:183–210, 1983. DOI: 10.1146/annurev.bb.12.060183.001151. 19

[17] Dill, K. A. Theory for the folding and stability of globular proteins, *Biochemistry*, 1985. DOI: 10.1021/bi00327a032. 19

[18] Dill, K. A., Bromberg, S., Yue, K., Chan, H. S., Ftebig, K. M., Yee, D. P., and Thomas, P. D. Principles of protein folding—a perspective from simple exact models, *Protein Sci.*, 4:561–602, 1995. DOI: 10.1002/pro.5560040401. 19

[19] Shirts, M. and Pande, V. S. Screen savers of the world unite!, *Science*, 290:1903–1904, 2000. DOI: 10.1126/science.290.5498.1903. 19

[20] Dobson, C. M., Swoboda, B. E. P., Joniau, M., and Weissman, C. The structural basis of protein folding and its links with human disease, *Philos. Trans. R. Soc. B Biol. Sci.*, 356:133–145, 2001. DOI: 10.1098/rstb.2000.0758. 19

[21] Berman, H. M., Westbrook, J., Feng, Z., Gilliland, G., Bhat, T. N., Weissig, H., and Shindyalov, I. N. The protein data bank, *Nucleic Acids Res.*, 28:235–242, 2000. DOI: 10.1093/nar/28.1.235. 21, 45

[22] PDB Current Holdings Breakdown. 21

[23] Karplus, M. and Kuriyan, J. Molecular dynamics and protein function, *Proc. Natl. Acad. Sci.*, 102:6679–6685, 2005. DOI: 10.1073/pnas.0408930102. 21, 22

[24] Hospital, A., Goñi, J. R., Orozco, M., and Gelpí, J. L. Molecular dynamics simulations: Advances and applications, *Adv. Appl. Bioinforma. Chem.*, 8:37–47, 2015. DOI: 10.2147/aabc.s70333. 21, 22

[25] McCammon, J. A., Gelin, B. R., and Karplus, M. Dynamics of folded proteins, *Nature*, 267:585–590, 1977. DOI: 10.1201/9781420035070. 21

[26] Levitt, M. and Sharon, R. Accurate simulation of protein dynamics in solution, *Proc. Natl. Acad. Sci.*, 85:7557–7561, 1988. DOI: 10.1073/pnas.85.20.7557. 21

[27] Noguti, T., Go, N., Ool, T., and Nishikawa, K. Monte Carlo simulation study of thermal fluctuations and conformational energy surface of a small protein, basic pancreatic trypsin inhibitor, *Biochim. Biophys. Acta—Protein Struct.*, 671:93–98, 1981. DOI: 10.1016/0005-2795(81)90098-2. 22

[28] Noguti, T. and Go, N. Collective variable description of small amplitude conformational fluctuations in a globular protein, *Nature*, 296:776–778, 1982. DOI: 10.1038/296776a0. 22

[29] Bahar, I. and Cui, Q. *Normal Mode Analysis: Theory and Applications to Biological and Chemical Systems*, Chapman & Hall, London, 2006. 22

[30] Dykeman, E. C. and Sankey, O. F. Normal mode analysis and applications in biological physics, *J. Phys. Condens. Matter*, 22:423202, 2010. DOI: 10.1088/0953-8984/22/42/423202. 22, 23

[31] Brooks, B. and Karplus, M. Harmonic dynamics of proteins: Normal modes and fluctuations in bovine pancreatic trypsin inhibitor, *Proc. Natl. Acad. Sci.*, 80:6571–6575, 1983. DOI: 10.1073/pnas.80.21.6571. 23, 24, 25

[32] Go, N., Noguti, T., and Nishikawa, T. Dynamics of a small globular protein in terms of low-frequency vibrational modes, *Proc. Natl. Acad. Sci.*, 80:3696–3700, 1983. DOI: 10.1073/pnas.80.12.3696. 23, 24, 25

[33] Levitt, M., Sander, C., and Stern, P. S. Protein normal-mode dynamics: Trypsin inhibitor, crambin, ribonuclease, and lysozyme, *J. Mol. Biol.*, 181:423–447, 1985. DOI: 10.1016/0022-2836(85)90230-x. 25, 26, 31, 39

[34] Ben-Avraham, D. Vibrational normal-mode spectrum of globular proteins, *Phys. Rev. B*, 47:14559–14560, 1993. DOI: 10.1103/physrevb.47.14559. 25, 26, 27

[35] Tirion, M. M. Large amplitude elastic motions in proteins from a single-parameter, atomic analysis, *Phys. Rev. Lett.*, 77:1905–1908, 1996. DOI: 10.1103/physrevlett.77.1905. 27, 28, 29

[36] Levitt, M. Molecular dynamics of native protein: I. Computer simulation of trajectories, *J. Mol. Biol.*, 168:595–617, 1983. DOI: 10.1016/S0022-2836(83)80304-0. 27

[37] Tozzini, V. Coarse-grained models for proteins, *Curr. Opin. Struct. Biol.*, 15:144–150, 2005. DOI: 10.1016/j.sbi.2005.02.005. 28, 29

[38] Rader, A. J. Coarse-grained models: Getting more with less, *Curr. Opin. Pharmacol.*, 10:753–759, 2010. DOI: 10.1016/j.coph.2010.09.003. 29

[39] Bahar, I. and Rader, A. J. Coarse-grained normal mode analysis in structural biology, *Curr. Opin. Struct. Biol.*, 15:586–592, 2005. DOI: 10.1016/j.sbi.2005.08.007.

[40] Takada, S. Coarse-grained molecular simulations of large biomolecules, *Curr. Opin. Struct. Biol.*, 22:130–137, 2012. DOI: 10.1016/j.sbi.2012.01.010. 29

[41] Bahar, I., Atilgan, A. R., and Erman, B. Direct evaluation of thermal fluctuations in proteins using a single-parameter harmonic potential, *Fold. Des.*, 2:173–181, 1997. DOI: 10.1016/s1359-0278(97)00024-2. 29, 30, 31

[42] Haliloglu, T., Bahar, I., and Erman, B. Gaussian dynamics of folded proteins, *Phys. Rev. Lett.*, 79:3090–3093, 1997. DOI: 10.1103/physrevlett.79.3090. 29, 30

[43] Bahar, I. and Jernigan, R. L. Cooperative fluctuations and subunit communication in tryptophan synthase, *Biochemistry*, 38:3478–3490, 1999. DOI: 10.1021/bi982697v. 31

[44] Bahar, I., Erman, B., Jernigan, R. L., Atilgan, A. R., and Covell, D. G. Collective motions in HIV-1 reverse transcriptase: Examination of flexibility and enzyme function, *J. Mol. Biol.*, 285:1023–1037, 1999. DOI: 10.1006/jmbi.1998.2371. 31

[45] Micheletti, C., Carloni, P., and Maritan, A. Accurate and efficient description of protein vibrational dynamics: Comparing molecular dynamics and Gaussian models, *Proteins Struct. Funct. Genet.*, 55:635–645, 2004. DOI: 10.1002/prot.20049. 31

[46] Yang, L., Song, G., and Jernigan, R. L. Protein elastic network models and the ranges of cooperativity, *Proc. Natl. Acad. Sci.*, 106:12347–12352, 2009. DOI: 10.1073/pnas.0902159106. 31, 37

[47] Zhang, H. and Kurgan, L. Sequence-based Gaussian network model for protein dynamics, *Bioinformatics*, 30:497–505, 2014. DOI: 10.1093/bioinformatics/btt716. 31

[48] Zhang, H., Jiang, T., Shan, G., Xu, S., and Song, Y. Gaussian network model can be enhanced by combining solvent accessibility in proteins, *Sci. Rep.*, 7:1–13, 2017. DOI: 10.1038/s41598-017-07677-9. 31

[49] Atilgan, A. R., Durell, S. R., Jernigan, R. L., Demirel, M. C., Keskin, O., and Bahar, I. Anisotropy of fluctuation dynamics of proteins with an elastic network model, *Biophys. J.*, 80:505–515, 2001. DOI: 10.1016/s0006-3495(01)76033-x. 32, 34, 35, 36, 37

[50] Eyal, E., Yang, L. W., and Bahar, I. Anisotropic network model: Systematic evaluation and a new web interface, *Bioinformatics*, 22:2619–2627, 2006. DOI: 10.1093/bioinformatics/btl448. 35, 37

[51] Carpinteri, A., Lacidogna, G., Piana, G., and Bassani, A. Terahertz mechanical vibrations in lysozyme: Raman spectroscopy vs. modal analysis, *J. Mol. Struct.*, 1139:222–230, 2017. DOI: 10.1016/j.molstruc.2017.02.099. 38, 39, 64, 65

[52] Carpinteri, A., Piana, G., Bassani, A., and Lacidogna, G. Terahertz vibration modes in Na/K-ATPase, *J. Biomol. Struct. Dyn.*, 37:256–264, 2019. DOI: 10.1080/07391102.2018.1425638. 38, 65

[53] Scaramozzino, D., Lacidogna, G., Piana, G., and Carpinteri, A. A finite-element-based coarse-grained model for global protein vibration, *Meccanica*, 54:1927–1940, 2019. DOI: 10.1007/s11012-019-01037-9. 38, 39, 40

[54] Giordani, G., Scaramozzino, D., Iturrioz, I., Lacidogna, G., and Carpinteri, A. Modal analysis of the lysozyme protein considering all-atom and coarse-grained finite element models, *Appl. Sci.*, 11:547, 2021. DOI: 10.3390/app11020547. 38, 39

[55] Hinsen, K. Analysis of domain motions by approximate normal mode calculations, *Proteins Struct. Funct. Genet.*, 429:417–429, 1998. DOI: 10.1002/(sici)1097-0134(19981115)33:3%3C417::aid-prot10%3E3.0.co;2-8. 40

[56] Hinsen, K., Thomas, A., and Field, M. J. Analysis of domain motions in large proteins, *Proteins Struct. Funct. Genet.*, 34:369–382, 1999. DOI: 10.1002/(sici)1097-0134(19990215)34:3%3C369::aid-prot9%3E3.0.co;2-f. 40

[57] Hinsen, K. and Kneller, G. R. A simplified force field for describing vibrational protein dynamics over the whole frequency range, *J. Chem. Phys.*, 11:10766–10769, 1999. DOI: 10.1063/1.480441. 40

[58] Tama, F., Gadea, F. X., Marques, O., and Sanejouand, Y. H. Building-block approach for determining low-frequency normal modes of macromolecules, *Proteins Struct. Funct. Genet.*, 41:1–7, 2000. DOI: 10.1002/1097-0134(20001001)41:1%3C1::aid-prot10%3E3.0.co;2-p. 41, 52

[59] Hoffmann, A. and Grudinin, S. NOLB: Nonlinear rigid block normal-mode analysis method, *J. Chem. Theory Comput.*, 13:2123–2134, 2017. DOI: 10.1021/acs.jctc.7b00197. 41, 52, 53

[60] Khade, P. M., Scaramozzino, D., Kumar, A., Lacidogna, G., Carpinteri, A., and Jernigan, R. L. hdANM: A new, comprehensive dynamics model for protein hinges, *Bioph. J.*, 120:4955–4965, 2021. DOI: 10.1016/j.bpj.2021.10.017. 41, 52, 53, 55, 56

[61] Tama, F. and Sanejouand, Y. H. Conformational change of proteins arising from normal mode calculations, *Protein Eng.*, 14:1–6, 2001. DOI: 10.1093/protein/14.1.1. 46, 48, 49, 50

[62] Harrison, R. W. Variational calculation of the normal modes of a large macromolecule: Methods and some initial results, *Biopolymers*, 23:2943–2949, 1984. DOI: 10.1002/bip.360231216. 47

[63] Brooks, B. and Karplus, M. Normal modes for specific motions of macromolecules: Application to the hinge-bending mode of lysozyme, *Proc. Natl. Acad. Sci.*, 82:4995–4999, 1985. DOI: 10.1073/pnas.82.15.4995. 47

[64] Marques, O. and Sanejouand, Y. H. Hinge-bending motion in citrate synthase arising from normal mode calculations, *Proteins Struct. Funct. Genet.*, 23:557–560, 1995. DOI: 10.1002/prot.340230410. 47

[65] Perahia, D. and Mouawad, L. Computation of low-frequency normal modes in macromolecules: Improvements to the method of diagonalization in a mixed basis and application to hemoglobin, *Comput. Chem.*, 19:241–246, 1995. DOI: 10.1016/0097-8485(95)00011-g. 48, 49

[66] Mahajan, S. and Sanejouand, Y. H. On the relationship between low-frequency normal modes and the large-scale conformational changes of proteins, *Arch. Biochem. Biophys.*, 567:59–65, 2015. DOI: 10.1016/j.abb.2014.12.020. 50, 51

[67] Delarue, M. and Sanejouand, Y. H. Simplified normal mode analysis of conformational transitions in DNA-dependent polymerases: The elastic network model, *J. Mol. Biol.*, 320:1011–1024, 2002. DOI: 10.1016/s0022-2836(02)00562-4. 50

[68] Krebs, W. G., Alexandrov, V., Wilson, C. A., Echols, N., Yu, H., and Gerstein, M. Normal mode analysis of macromolecular motions in a database framework: Developing mode concentration as a useful classifying statistic, *Proteins Struct. Funct. Genet.*, 48:682–695, 2002. DOI: 10.1002/prot.10168. 50, 51

[69] Tobi, D. and Bahar, I. Structural changes involved in protein binding correlate with intrinsic motions of proteins in the unbound state, *Proc. Natl. Acad. Sci.*, 102:18908–18913, 2005. DOI: 10.1073/pnas.0507603102. 51

[70] Dobbins, S. E., Lesk, V. I., and Sternberg, M. J. E. Insights into protein flexibility: The relationship between normal modes and conformational change upon protein-protein docking, *Proc. Natl. Acad. Sci.*, 105:10390–10395, 2008. DOI: 10.1073/pnas.0802496105. 51

[71] Nicolay, S. and Sanejouand, Y. H. Functional modes of proteins are among the most robust, *Phys. Rev. Lett.*, 96:1–4, 2006. DOI: 10.1103/physrevlett.96.078104. 51

[72] Mahajan, S. and Sanejouand, Y. H. Jumping between protein conformers using normal modes, *J. Comput. Chem.*, 38:1622–1630, 2017. DOI: 10.1002/jcc.24803. 51

[73] Yang, L., Song, G., and Jernigan, R. L. How well can we understand large-scale protein motions using normal modes of elastic network models?, *Biophys. J.*, 93:920–929, 2007. DOI: 10.1529/biophysj.106.095927. 51

[74] Gerstein, M. and Krebs, W. A database of macromolecular motions, *Nucleic Acids Res.*, 26:4280–4290, 1998. DOI: 10.1093/nar/26.18.4280. 52

[75] Khade, P. M., Kumar, A., and Jernigan, R. L. Characterizing and predicting protein hinges for mechanistic insight, *J. Mol. Biol.*, 432:508–522, 2020. DOI: 10.1016/j.jmb.2019.11.018. 53, 54

[76] Raman, C. V. A new radiation, *Proc. Ind. Acad. Sci.—Sect. A*, 37:333–341, 1953. DOI: 10.1007/bf03052651. 59

[77] Thomas, G. J. Raman spectroscopy of protein and nucleic acid assemblies, *Annu. Rev. Biophys. Biomol. Struct.*, 28:1–27, 1999. DOI: 10.1146/annurev.biophys.28.1.1. 62

[78] Rygula, A., Majzner, K., Marzec, K. M., Kaczor, A., Pilarczyk, M., and Baranska, M. Raman spectroscopy of proteins: A review, *J. Raman Spectrosc.*, 44:1061–1076, 2013. DOI: 10.1002/jrs.4335. 62

[79] Bunaciu, A. A., Aboul-Enein, H. Y., and Hoang, V. D. Raman spectroscopy for protein analysis, *Appl. Spectrosc. Rev.*, 50:377–386, 2015. DOI: 10.1080/05704928.2014.990463. 62

[80] Kuhar, N., Sil, S., Verma, T., and Umapathy, S. Challenges in application of Raman spectroscopy to biology and materials, *RSC Adv.*, 8:25888–25908, 2018. DOI: 10.1039/c8ra04491k. 62

[81] Lacidogna, G., Piana, G., Bassani, A., and Carpinteri, A. Raman spectroscopy of Na/K-ATPase with special focus on low frequency vibrations, *Vib. Spectrosc.*, 92:298–301, 2017. DOI: 10.1016/j.vibspec.2017.08.002. 62, 63, 64

[82] Brown, K. G., Erfurth, S. C., Small, E. W., and Peticolas, W. L. Conformationally dependent low-frequency motions of proteins by laser Raman spectroscopy, *Proc. Natl. Acad. Sci.*, 69:1467–1469, 1972. DOI: 10.1073/pnas.69.6.1467. 62, 63

[83] Genzel, L., Keilmann, F., Martin, T. P., Wintreling, G., Yacoby, Y., Fröhlich, H., and Makinen, M. W. Low-frequency Raman spectra of lysozyme, *Biopolymers*, 15:219–225, 1976. DOI: 10.1002/bip.1976.360150115. 63

[84] Painter, P. C., Mosher, L., and Rhoads, C. Low-frequency modes in the Raman spectra of proteins, *Biopolymers*, 21:1469–1472, 1982. DOI: 10.1002/bip.360210715. 63, 64

[85] Kalanoor, B. S., Ronen, M., Oren, Z., Gerber, D., and Tischler, Y. R. New method to study the vibrational modes of biomolecules in the terahertz range based on a single-stage Raman spectrometer, *ACS Omega*, 2:1232–1240, 2017. DOI: 10.1021/acsomega.6b00547. 63

[86] Neu, J. and Schmuttenmaer, C. A. Tutorial: An introduction to terahertz time domain spectroscopy (THz-TDS), *J. Appl. Phys.*, 124:231101, 2018. DOI: 10.1063/1.5047659. 66, 67

[87] Falconer, R. J. and Markelz, A. G. Terahertz spectroscopic analysis of peptides and proteins, *J. Infrared, Millimeter, Terahertz Waves*, 33:973–988, 2012. DOI: 10.1007/s10762-012-9915-9. 68

[88] Markelz, A. G., Roitberg, A., and Heilweil, E. J. Pulsed terahertz spectroscopy of DNA, bovine serum albumin and collagen between 0.1 and 2.0 THz, *Chem. Phys. Lett.*, 320:42–48, 2000. DOI: 10.1016/s0009-2614(00)00227-x. 68

[89] Markelz, A., Whitmire, S., Hillebrecht, J., and Birge, R. THz time domain spectroscopy of biomolecular conformational modes, *Phys. Med. Biol.*, 47:3797–3805, 2002. DOI: 10.1088/0031-9155/47/21/318. 68, 69, 70

[90] Chen, J. Y., Knab, J. R., Ye, S., He, Y., and Markelz, A. G. Terahertz dielectric assay of solution phase protein binding, *Appl. Phys. Lett.*, 90:1–4, 2007. DOI: 10.1063/1.2748852. 68, 72

[91] Whitmire, S. E., Wolpert, D., Markelz, A. G., Hillebrecht, J. R., Galan, J., and Birge, R. R. Protein flexibility and conformational state: A comparison of collective vibrational modes of wild-type and D96N bacteriorhodopsin, *Biophys. J.*, 85:1269–1277, 2003. DOI: 10.1016/s0006-3495(03)74562-7. 69

[92] Balu, R., Zhang, H., Zukowski, E., Chen, J. Y., Markelz, A. G., and Gregurick, S. K. Terahertz spectroscopy of bacteriorhodopsin and rhodopsin: Similarities and differences, *Biophys. J.*, 94:3217–3226, 2008. DOI: 10.1529/biophysj.107.105163. 69, 70

[93] Acbas, G., Niessen, K. A., Snell, E. H., and Markelz, A. G. Optical measurements of long-range protein vibrations, *Nat. Commun.*, 5:1–7, 2014. DOI: 10.1038/ncomms4076. 70, 71

[94] Niessen, K. A., Xu, M., and Markelz, A. G. Terahertz optical measurements of correlated motions with possible allosteric function, *Biophys. Rev.*, 7:201–216, 2015. DOI: 10.1007/s12551-015-0168-4. 71

[95] Niessen, K. A., Xu, M., Paciaroni, A., Orecchini, A., Snell, E. H., and Markelz, A. G. Moving in the right direction: Protein vibrational steering function, *Biophys. J.*, 112:933–942, 2017. DOI: 10.1016/j.bpj.2016.12.049. 71, 72

Authors' Biographies

DOMENICO SCARAMOZZINO

Domenico Scaramozzino received his B.Sc. and M.Sc. in Civil Engineering from Politecnico di Torino in Italy. He was recently awarded his Ph.D. in Civil and Environmental Engineering from the same university, where he is now working as a Postdoctoral research fellow in Structural Mechanics. His research interests are mainly focused on Elastic Lattice Models, which are applied both in the traditional fields of Structural Engineering, e.g., for the analysis of trussed shell structures and tall buildings, as well as in the most advanced fields of Computational Biology and Bioinformatics, with special focus on proteins and macromolecules. Related to this last topic, he was also a Visiting Scholar in the Department of Biochemistry, Biophysics and Molecular Biology at Iowa State University during the academic year 2019–2020.

GIUSEPPE LACIDOGNA

Giuseppe Lacidogna received his Ph.D. in Structural Engineering from Politecnico di Torino in 1994, where he graduated cum Laude in Architecture. He is an Associate Professor in Structural Mechanics at Politecnico di Torino and achieved the National Academic Qualification as Full Professor of Structural Mechanics since 2018. He is an Officer of the European Academy of Sciences (EurASc), member of the Editorial Board of several international journals, and affiliated to different scientific associations (among them AIMETA, IA-FraMCoS, SEM, RILEM, and ISA). He is the author of more than 120 papers in refereed international journals, 24 book chapters, and 7 monographs. In 2018, he received a Certificate Merit Award from the European Society for Experimental Mechanics (EuraSEM), and he is in the list of "Top Italian Scientists."

ALBERTO CARPINTERI

Alberto Carpinteri received his Doctoral Degrees in Nuclear Engineering cum Laude (1976) and in Mathematics cum Laude (1981) from the University of Bologna in Italy. He is currently Chair Professor of Solid and Structural Mechanics at Politecnico di Torino (1986–), Head of the Engineering Division in the European Academy of Sciences (2016–), Honorary Professor at Tianjin University (2017–), and Zhujiang (Pearl River) Professor of Guangdong Province, Shantou University (2019–). He was the President of different Scientific Associations and Research Institutions: the International Congress on Fracture, ICF (2009–2013), the European Structural Integrity Society, ESIS (2002–2006), the International Association of Fracture Mechanics for Concrete and Concrete Structures, lA-FraMCoS (2004–2007), the Italian Group of Fracture, IGF (1998–2005), the National Research Institute of Metrology, INRIM (2011–2013). He is the author or editor of over 900 publications, of which more than 400 are papers in refereed international journals and 55 are books or journal special issues. He received numerous international Honours and Recognitions: the Robert L'Hermite Medal from RILEM (1982), the Griffith Medal from ESIS (2008), the Swedlow Memorial Lecture Award from ASTM (2011), the Inaugural Paul Paris Gold Medal from ICF (2013), the Doctorate Honoris Causa in Engineering from the Russian Academy of Sciences (2016), and the Frocht Award from SEM (2017).

Printed in the United States
by Baker & Taylor Publisher Services

Printed in the United States
by Baker & Taylor Publisher Services